POLICIES & PROCEDURES FOR

DATA SECURITY

A Complete Manual for Computer Systems and Networks

BY THOMAS R. PELTIER

A Computer Security Institute Book

Miller Freeman Books

San Francisco

To Judy & Justin and Jill & Tom

À *Computer Security Institute* Book
Miller Freeman Inc., 600 Harrison St., San Francisco, CA 94107

ISBN 0-87930-239-9
Library of Congress Catalog Card Number 91-61853
Printed in the United States of America

94 95   5 4 3

CONTENTS

PREFACE

This book was written for all security officers and data processing managers charged with developing policies and procedures on information and computer security. The book includes numerous checklists and examples to help the reader create a security program to suit the style and requirements of the reader's own organization.

The premise of this book is that data processing security procedures are necessary in all computerized organizations. While no one would identify security as a company's most important product, security is nevertheless necessary for a company to carry on its business, whether that involves manufacturing a product or providing a service. Any decision about what controls are necessary for the security of the organization must be tempered by the reality of the company's business needs.

The goal of this book is to provide the practical knowledge to identify the need for security, research existing measures, and formulate an effective program that will be supported by all members of the corporation, from senior managers to end-users.

The cornerstone of an effective security program is the corporate policy statement. The policy statement forms the basis for all successive company guidelines, standards, regulations, and procedures. An information security policy statement must identity company assets, such as information and systems. The policy statement must also define who is responsible for classifying the assets, what role the employees play in protecting the assets, where the policy is enforced, and who is responsible for monitoring compliance with the policy.

Once information and data have been recognized as company resources, a data classification scheme is essential. This book outlines the criteria used to legally determine if information is confidential, provides definitions and examples of various data classification levels, describes the principles used to ensure data integrity, and discusses the role of management and of the information owner and custodian in the care and control of information. The book also describes copyright restrictions and methods for declassification and destruction of sensitive material.

Like the policy statement, the mission statement is an important part of the security program because it provides a kind of job description for the information security department. The mission statement defines the scope and direction of information security within the company. Typical responsibilities that are spelled out in the mission statement include recommending security policies and procedures, evaluating new technology and recommending security strategies to protect it, monitoring changes in laws and regulations that affect security requirements, and developing a security awareness program for the organization.

Once the mission statement is endorsed, the specific security procedures for each area of the organization can be developed, such as physical security, security administration, application development, system programming, and telecommunications. As different areas of the security manual are written, a review panel will be needed to evaluate each draft and provide written comments on the process.

To be effective, not just in theory but in reality, all procedures should have the support of management as well as the end-users of the information. To gain this support, information security officers will have to educate management and employees about the benefits of the program and justify the cost. The book gives many proven techniques for gaining support for security programs and for developing an organization-wide employee awareness program. The book also includes a chapter on applicable computer and security information laws for the United States and for other major countries.

CHAPTER ONE

Policy Statement

THE CORNERSTONE OF ALL PROCEDURES

Every data processing installation has the responsibility for defining and publishing a set of security standards. Many corporations are required by law to set standards. Through the Foreign Corrupt Practices Act (FCPA) of 1977, publicly held companies are required to "establish policies and procedures to safeguard asset loss and produce reliable financial records." All federal agencies, many government contractors, and some state agencies have been mandated by the Computer Security Act (CSA) of 1987 to:

- Develop guidelines for protecting unclassified, but sensitive, information stored in government computers.
- Formulate a computer security plan.
- Train employees on the threats to and vulnerabilities of their computer systems.

Each agency or contractor must come to a full understanding of the CSA provisions and apply them as required.

For organizations that are not subject to either the FCPA or CSA regulations, proper controls and continuation of service are spelled out in contracts with their customers. All organizations in the public, private, or government sector are therefore required to establish effective computer security policies and procedures either by law, contract, or just plain good business practice.

Most computer security policies and procedures are based on the realization that the various computer systems are completely integrated into the organization. Managers are beginning to understand that it is their responsibility to ensure that a system of internal controls has been established and maintained. This system provides reasonable assurances that the books and records reflect the transactions of the organization and that its established policies and procedures are carefully followed. Failure in any of these areas can be viewed as a breach of senior management's fiduciary responsibility.

When researching the material for this book, I reviewed policy statements from the early 1950s to the present. Through the stated objectives, it was clear that each decade brought its own set of concerns to the data processing, computer operations, information systems security, and, now, the corporate information office.

Just prior to my introduction to computers in 1966, the key objective was keeping the system up. Security procedures and controls were not really necessary. Few people could actually code in the primitive languages of machine and/or Assembler. In the mid 60s, IBM introduced the third generation of computers—the 360 series. These were powerful machines and organizations began to increase their data processing staffs, as computer labs became the showcase of progressive-thinking management.

With the threats of violence in the late 60s and early 70s, the computer security industry began to develop. Controls in the 70s were physical in nature, resulting in the hidden computer rooms with access card readers, Halon, offsite tape storage, and disaster recovery plans. During this time, interactive data processing came into its own, and passwords were replaced by access control packages such as RACF, SAC, and ACF2. More and more people were able to program, although programming was still restricted to those who knew COBOL and PL1.

In 1983, the rise of the affordable microcomputer gave users control of their computing environment. For too long a period, the user community had been held hostage by the data processing departments. Now the users had computers, and the number of computers increased dramatically. In 1985, William Webster, head of the FBI, told the Computer Security Institute (CSI) Annual Conference that there were 3.3 million personal computers worldwide, with most of them concentrated in the financial industry. At the 1990 CSI IBM/DEC conference, it was estimated that there were 58 million personal computers used for all types of business, education, and personal processing. As computers became common in everyday businesses, connectivity became the issue of the 1980s. Coupled with the ever-increasing development of universal access networks, maintaining the integrity of information, data, programs, and systems is quickly becoming the issue of the 1990s.

With computer use increasing around the world, the need for organization-wide policy statements becomes more apparent. The goal must be to establish a universal message that all employees, from the entry level to top management, can understand and embrace. Additionally, without senior management's understanding and support of the objectives of an information security program, its ultimate success is tenuous.

Developing computer security policies and procedures is part of an overall information security program. The goal of an organization-wide policy statement is to lay the cornerstone for the security structure. The policy statement will support the infrastructure of all other charters, guidelines, practices, procedures, and orders for the company.

Before examining existing policy statements and creating a policy statement of your own, some common terminology must be discussed.

When creating a policy statement for an organization, the terms "data" or "information" often become confused. To many readers, the term "data" refers to or has been linked to data processing. There is, however, a difference between data and information, and the differences are pivotal in providing accountability and responsibility. As Robert H. Courtney, Jr., president of RCI, put it "Information is a subset of data that has been selected to answer a specific need. Thus the difference between data and information is precisely the difference between the content of the telephone directory (which is data) and the number you happen to want at the time (which is information)."[1] Because of this distinction, both data and information must be controlled. Data is data processing and, for most users, is equated with DP staffs. When creating policy statements, you should avoid the narrow interpretation of the term "data" and address the broad issue of information protection. The chapters on charters and mission statements will incorporate the issue of data protection. Figure 1 shows the typical hierarchy of security control.

The basis of security control is provided by the laws, contracts, IRS procedures and rulings, banking circulars, and regulations that affect your organization. These regulations are established to ensure that business is conducted in an orderly manner, that controls are in place to ensure that information is protected from unauthorized access, modification, disclosure and/or destruction,

FIGURE I **PROGRESSION OF SECURITY CONTROL**

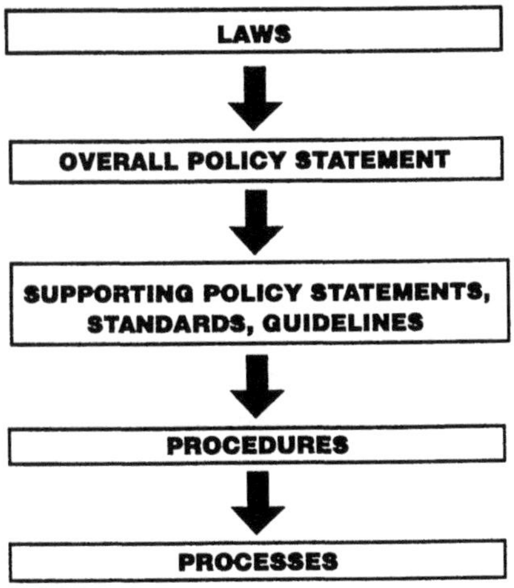

that transactions are authorized, and that sufficient audit trails have been generated to allow for adequate review and reconciliation.

INFORMATION AS AN ASSET

An effective policy statement provides the basis for all other control documents. To create an effective policy statement, you must recognize that information is an asset of the organization. For the past several years, public and government documents have reinforced the idea that information is a valuable corporate asset and that senior management is responsible for controlling that information.

In the proceedings that led to the Computer Fraud and Abuse Act of 1986, Senator Strom Thurmond, from the Judiciary Committee, stated that "Enforcement problems can largely be overcome by recognizing computerized information as property. The Congress began that recognition by enacting the Computer Fraud Abuse Act of 1984. The Committee intends S.2281 to affirm the government's recognition of computerized information as property."[2]

Thomas R. Horton, president and CEO of American Management Association, endorsed the concept when he said that "In any business environment today, one of the most valuable assets is information. Operations data, research data, financial data . . . all are critical to today's manager, and all must be controlled. At the same time, we must realize that there is little difference between managing information and managing anything else. The best-managed companies have always known this."[3]

Computers have become common throughout most organizations. The modern office contains at least a terminal and often has either a companion personal computer or emulation adaptor. These intrusions are accepted because they are necessary to "get the job done." The desire to complete the tasks at hand must, however, be tempered by the organization's commitment to maintain a secure computing environment. Security can only be accomplished by publishing policies and procedures that are, in turn, supported by management at all levels.

- To be effective, a policy statement should speak to employees in their language. Whenever you prepare a statement for employees, you should make certain that you explain exactly what is expected. Don't try to be folksy, simply state what must be done and state it in terms that all employees can understand. A later chapter will discuss the development of an employee awareness program, but before employees can be made aware of their responsibilities, the general direction must be defined in the policy statement.

Ray J. Groves, former chairman and chief executive of Ernst and Whinney wrote that "Computer systems are an important corporate asset that must not be misused. Crucial to protecting this asset is corporate management's understanding that controlling computer systems requires a perspective that extends beyond the computer room."[4] With today's decentralized environment, this may seem apparent to the computer security professional; however, many senior managers believe that if the computer room has been locked up, then everything must be secure. Mr. Groves further states that "No internal control system can provide absolute assurance that all errors and irregularities . . . will be prevented or detected. But by looking beyond the computer room, and continually assessing the adequacy of internal controls over computer-generated data, corporate management—with oversight and review . . . can minimize the risk."[5]

Mr. Groves addresses three very important security issues. First, all computer systems, not just the hardware, are an important corporate asset. Second, management must be made aware of the expanding importance of computer-generated data and information; this means that controls must move beyond the walls of the computer room. Third, no matter how effective and comprehensive, no set of policies and procedures can prevent or detect all irregularities.

MANAGEMENT RESPONSIBILITY

As each functional area of a company (such as payroll, personnel, and purchasing) converted from manual processing to computer-based processing, a fundamental business control was muddled. For a number of reasons, the responsibility for protecting access to information was abdicated to the data security technicians. Only now is the responsibility reverting back to its proper location, with the users.

Robert P. Bigelow, an attorney who writes on computer-related issues, wrote that "Management, whether it realizes it or not, is responsible for the security of the organization's information system. Sometimes this responsibility is contractual, sometimes it is created by law, and sometimes it will be imposed by the courts. If top management does not act to establish and maintain adequate data security procedures, the organization could be liable for substantial damages . . . and those damages may be collectible from the managers personally!"[6]

In today's decentralized environment, there has been a reemergence of the functional information owner. In a security context, this means that the department that created the data or the primary user of the data is responsible for setting the rules for information access and usage. Security is not a data processing responsibility, but lies with the managers of engineering, personnel, manufacturing, purchasing, payroll, accounts payable, or scheduling.

As operations and information processing become more decentralized, the need for established controls and procedures has increased. As information moves away from the confines and controls of the main computer facility, the potential for unauthorized access, modification, disclosure, or destruction will increase. Clearly, in this environment, the ultimate responsibility for information protection lies not with the provider of the service, but with the department manager and, finally, the top executives. The responsibility for protecting information assets lies with management, regardless of where the information is stored—in a mainframe computer, a personal computer, a minicomputer, a desk, a file cabinet, a waste basket, or a human mind. In a mainframe environment, the data processing staff will assist the other departments in ensuring that the data is properly protected, but the responsibility still remains with the individual departments. Larry Horner stated the issue succinctly: "As operations and related information processing become more decentralized, controls and procedures become more important. And as remote access, networks, and distributed data processing make possible greater globalization and decentralization, the potential for unauthorized access to confidential and proprietary information grows apace. As the ultimate bearer of corporate responsibility . . . it's up to the CEO to make sure that appropriate computer security mechanisms are in place."[7]

DECENTRALIZATION AND CONNECTIVITY

As discussed, the number of microcomputers grew from 3.3 million in 1985 to nearly 60 million today. Even more alarming is the increasing number of people with direct access to computers. In 1988, Patrick J. McGovern, chairman of the International Data Group, explained that ". . . there are 50 million people with direct access to computers . . . By 1992, some 350 million people will be working with computers; and as we enter the next century, the computing population will reach 2.4 billion."[8] Recent research on this issue suggests that the estimates may be low. With the French telephone companies giving free terminals and modems to subscribers (and with test marketing of a similar system in the United States soon), there may be an even larger audience wanting access to computer systems and the information in them. With such potential access, the need for control becomes even more important for every organization.

From the preceding discussion, the need for a written policy statement is certainly clear. It is now time to begin writing a policy statement for your organization. As you write a policy statement, you will define how information and computer systems are essential to the business of your organization. The goal of the policy statement is to establish a clear corporate direction for security. Additionally, the policy statement should recognize that all employees are integral to the security process and that senior management should be supported in fulfilling its responsibility for protecting the information assets of the company.

UNDERSTANDING THE SITUATION

The key to developing an effective policy statement is to understand who will be affected by the policy. Each organization is different. A financial institution has a different set of objectives and needs than a research laboratory; both require direction to ensure that assets are properly protected. You need to understand the specifics of your organization to ensure that the policy statement you create will meet your objectives. Only you know what the problems of your own organization are, and you will use that background to develop standards and procedures to address the problems.

It might be tempting to thumb ahead a few pages and copy out one of the sample policy statements. But each statement was developed to handle the specific needs of an organization. You should never create a policy first and then use it to define the level of security needed. It is far more effective to define the security needs of your organization and create a policy statement to satisfy them.

Sometimes, when you work in an organization, you get a very myopic view of it. After all, you are doing your job and generally only interact with a limited number of employees. Before beginning the process of policy development, you will have to broaden your knowledge of what the organization does and consider the security requirements of the entire company.

Recent audit reports relating to computer and information security are a great source of information for determining the problems within your organization. It is important to go back at least five years and to look at comments outside the data processing department. You should review the internal audit staff reports as well as the external auditors' findings.

At this time, you should also conduct a risk assessment or analysis. A risk assessment will allow the organization to identify assets, threats to those assets, and potential controls. Be sure to enlist other departments in this process. Remember, the policy statement will be your first attempt at establishing credibility within your organization. Take the time to do the job right and determine what needs to be protected and how the organization views the need for security.

WRITING A POLICY STATEMENT

Policies are the general plans made by management for providing information and direction. Policies establish the basic philosophy of the organization and determine the areas where controls must be established.

To be effective, an information security policy statement must identify information and systems as a company asset and identify who is responsible for classifying the assets and what role the employees play in the overall structure of control and protection.

ESTABLISHING THE POLICY SCOPE

Employees are any organization's most important asset. Just as steps have been taken to ensure that employees are protected within the work environment, their support must now be enlisted to protect the organization's second most important asset—classified information. If the organization and the employees fail to protect the information from unauthorized access, the organization will lose customer confidence, competitive advantage, and, ultimately, jobs.

The audience must be identified in the policy statement, and generally the audience includes all regular, full-time, salaried employees. You then must decide if the policy statement will apply to contract, co-op, summer interns, per diem, part-time, and hourly employees. Will the policy statement also apply to anyone who handles the organization's information? If you include hourly workers in the scope of the policy, you may need to get union approval if the hourly employees are represented by one.

Chapter 5 will discuss how to establish a review panel, but you should involve the labor relations department in your development as soon as possible since the labor relations department will be helpful in gaining acceptance of your overall program. The labor relations staff may be able to offer alternatives for the policy. For example, most union contracts have established a set of shop rules that govern the day-to-day activities of the represented employees. There may already be a shop rule in effect that will provide the level of security required, and you may only need to reference the specific shop rule at the appropriate time and place.

When defining the audience for your policy, make sure the policy statement stipulates all of the employees you wish to cover. In any organization, there are a few "wannabees" who aspire to be lawyers, auditors, or loophole finders. The "wannabees" read everything that is published and find the fastest way around the policy; they want to be sure that any new policy won't apply to them or their department. While policies are being developed to support the overall organization, the "wannabees" often do not see, understand, or even care about the big picture. So you have to make it clear that the policy you create applies to everyone in your organization, and leaves no doubt in anyone's mind that they are affected by the contents.

While the straightforward approach is recommended, you will also have to temper the wording to suit your organization and draw on previously published works for guidelines on what to write. The following section includes two examples of usage and responsibility statements.

USAGE AND RESPONSIBILITY STATEMENT

EXAMPLE 1 **ACKNOWLEDGING RESPONSIBILITIES**

All managers, administrators, and users of automated information processing systems (all owned, leased, and contracted services involving word processors, micro and minicomputers, mainframes, and service bureaus) will be required to read and sign a form containing the following personal usage and responsibility statement:

I understand that unauthorized use of, or contribution to unauthorized use of, computer facilities and/or corporate data constitutes a major infraction. I recognize that I am responsible for maintaining the confidentiality of corporate information that I access while employed by the company. Failure to comply with these responsibilities may result in suspension without pay or immediate discharge without prior warning.

EXAMPLE 2 **AUTHORIZATION FOR ACCESS**

All managers, administrators, and users of computer systems requesting new system access, or those users requesting a change in existing status, will be required to read and sign a form containing the following personal usage and responsibility statement:

I understand that unauthorized use of or contribution to unauthorized use of computer facilities and/or corporate data constitutes a violation of corporate policies. I recognize that I am responsible for maintaining the confidentiality of corporate information that I have access to while employed by the corporation and that failure to comply with these responsibilities is considered a violation of the Employee Code of Conduct.

Both examples of usage and responsibility statements basically say the same thing. The first example was created by a bank. While I found nothing wrong with the statement, a number of employees objected to signing a document that stated, "Failure to comply with these responsibilities may result in suspension without pay or immediate discharge without prior warning."

The statement was more appropriate for a bank than for my organization. I had to be diplomatic. To make the usage and responsibility statement less threatening to my employees, it was reworded to state, "Failure to comply with these responsibilities is considered a violation of the Employee Code of Conduct." The removal of the threat of suspension made the statement acceptable to the employees. For the record, the Employee Code of Conduct states that the failure to comply with company policies and procedures may result in severe disciplinary sanctions up to and including discharge without prior warning.

Both usage and responsibility statements are clear and easily understood by all employees; both were acceptable within their own organizations. Your responsibility is to understand how your organization expresses itself in written form and adopt that style into your policies and procedures.

ESTABLISHING THE PURPOSE

One key to writing an effective policy is to understand why policies are important. From a legal standpoint, an effective policy statement documents that your organization understands its responsibilities and is taking steps to safeguard its assets. The policy statement is your organization's declaration that protection of assets is a key goal of the organization. Remember, your senior managers have a fiduciary responsibility to protect the assets that have been entrusted to them. The policy statement identifies the assets and clearly states the organization's commitment to protect them.

When you are writing your procedure manual, each chapter should begin with a policy or overview statement to describe the contents of the chapter. You should review existing policy and procedure manuals within your organization to ascertain how your organization has defined its security measures to date.

Consistency is the key to acceptance, both internally and externally. When developing your policy statement, you should make the length and style similar to those found in other organizational documents, such as accounting, tax, operations, engineering, purchasing, and personnel manuals. Until you understand how your own organization puts forth these goals, you should not begin writing. Review of existing organization documents will help you establish the purpose of your policy statement.

DEVELOPING THE POLICY DIRECTION

In any discussion of policy development, two management styles are invariably mentioned. The first is the top–down method, in which the people in authority decree what the policy will be. The other method is the grass roots effort, in which the people affected by the policy review and discuss it. Both styles have their positive sides and their pitfalls.

Because the top-down method is usually supported by management, it has a better chance of success, but many times it lacks reality. When top management decrees what the policy is to be, the policy may be too general or, because top management is responsible for maintaining the assets of the organization, too restrictive.

The grass roots effort engages all the employees, but often gets bogged down in discussion or only addresses each individual's vested interests.

Your goal is a policy that combines the best of both methods: sort of a "top-roots" method. After you research your policy statement, you will have it reviewed by people in the field (the review process is discussed in depth in Chapter 5), and then you will get the support of top management. This procedure will provide you with the best policy and the highest level of support. The review process gives you employee input and acceptance, and the support of top management will ensure that the policy is implemented throughout the organization. In order for your policy to be effective, you will need both.

ESTABLISHING THE ENFORCEMENT CONCEPTS

Clear accountability for policy adherence must be established. All job descriptions should cover the security responsibilities of the employees. In addition, orientation for new employees should cover security responsibilities, and the existing staff should be briefed on new procedures. Chapter 7 will discuss the development of an employee awareness program, but the enforcement concepts are the backbone of the overall security program.

There must be a clear reason why the organization is requiring employees to abide by security policies. At annual reviews, the employee's adherence to the security requirements of the job should be evaluated as an indicator of employee performance. During reviews, employees should also review and confirm the usage and responsibility statements.

Reviews, awareness training, and enforcement are never-ending processes. If an employee is only subjected to the organization's policies on information and computer security at his or her annual review, the policy lacks meaning. The written word is only effective if the employees are aware of its existence and aware that their compliance with it is important to the company.

Ideally, the policy will be enforced equally throughout the organization. However, equal may mean different things at different levels of the organization. You will therefore need to consider the reality of business life and avoid a policy that is doomed to failure. Existing personnel department policies can provide a good source of information for structuring the definitions and enforcement requirements.

When personal computers were first introduced in the workplace, companies struggled with how to control their use. Initially, many companies took a hard line and wrote policies reserving personal computers for authorized business use only. This policy seemed like a good idea, but there was no way to enforce it. Furthermore, it quickly became apparent that the employees were going to use the computers for personal use, whether allowed to or not. After a short time, the policies had to be reviewed.

It was determined personal computer use should be similar to use of company telephones. The telephone on an employee's desk is there for business purposes only. However, there probably isn't a telephone within your company that hasn't been used to make a personal phone call today (except the pay phones in the lobby). Companies know that employees use the phones for personal calls, and most companies accept this as a part of doing business. However, the policy that evolved was to review phone bills regularly and have the employees reimburse the company for any out-of-the-ordinary use. With this policy in place, employees are aware of their responsibilities and realize that there is enforcement to back it up.

Using the policy on phone use as a guideline, the policy on personal computers was updated to state that "use of company-provided personal computers was to be limited to authorized business use only, unless authorized by local management." This new direction allowed the company to be a winner in several ways. By approving employees' use of computers on their own time, employees were now eager to explore and learn more about the personal computer—beyond what was necessary to just do their jobs.

In one instance, an employee created a Lotus spreadsheet to keep the company golf league standing and handicaps. During the annual budget process within the employee's department, his boss indicated that he needed a spreadsheet to perform certain calculations. The golf league employee was able to create the budget analysis spreadsheet using the skills he had learned while using the computer on his own time.

By allowing employees to use computers on their own time, the employees gained more mastery over the tools made available to them. Additionally, if the employees have personal items stored on the computer (such as letters, checkbook data, or a college research paper) then they have a vested interest in protecting data.

By listening to the employees, the policy was changed in a way that was beneficial to all parties. The goal was to ensure that the equipment was not being abused or being used to run an employee business. In the process, the company gained better-educated employees at no cost, and the security program gained employees that were sensitive to the issues of protecting data. In addition, abuse of the system was prevented. If the old policy had been continued, the computers would have been used for personal use anyway. By providing authorized and controlled use, employees were spared violating the policy and rationalizing why they did not conform to it.

When developing a policy, you will have to ensure that there are ways to measure enforcement. Enforcement cannot be left to the audit staff. The department managers must be actively involved in the review and control process. Many companies have established an internal control review program. This program consists of a series of questions on specific topics that are to be completed annually. The questions are broken down by functional areas—purchasing questions, purchasing answers, receiving questions, receiving answers, and so forth. However, managers in every department are required to answer the questions on information security. Sample questions are included in Chapter 8.

PRESENTING THE POLICY

The actual format of your policy will depend on the style of your organization. Before you begin writing the policy, you should take time to read some of the policy statements from other areas in your organization. Then determine which of the following two styles is right for your company.

Mandatory Requirements. The policy is stated in a specific manner and exceptions, if allowed, are specifically defined in the policy.

> *Example:* Information owners are the senior level managers that have been formally assigned the organization's proprietary rights and fiduciary responsibilities for the information. The information owner is responsible for the following:
> 1. Judging the value of the information and assigning a proper classification.
> 2. Authorizing access to the information, transaction, etc.
> 3. Specifying control requirements and communicating these requirements to the information custodian/steward.
>
> The information owner may delegate these responsibilities to another individual (for example, a department manager to a department employee); however, the responsibilities are not to be delegated to the steward or custodian.

Mandatory policy determines who is responsible and authorizes only a limited degree of delegation.

Minimum Requirements. This policy style establishes a level of control that all employees are expected to meet, while leaving room for variation beyond that level.

Actually, this type of policy is the most flexible and realistic. It sets a baseline for controls that will be used to measure locally developed procedures, similar to the relationship between the federal and state governments. The federal laws set the minimum standards, and the states then add further restrictions. For example, federal law requires that all drivers must be licensed by the state of their permanent residence, and state laws determine the age, training classes, and testing of each driver.

> *Example:* All employees are to be aware of the important responsibility that they have to safeguard all organizational information, and they are to be thoroughly instructed in sound security practices and procedures. This educational process should be part of the new employee orientation process and is to remain part of the employee's ongoing educational process.

The policy states that a security awareness program must be established and that all employees must be made aware of their responsibilities on a regular basis. The policy does not discuss training content, class length, frequency of training, training dates, or attendees. This policy leaves room for local variation in order to meet the needs of local employees.

WRITING THE POLICY

As stated previously, the written policy should clear up confusion, not generate it. The policy must clearly state the objectives and answer the following questions:

What is to be protected? State the obvious. Describe for your employees (and those who will review and audit your organization's controls) just what is important and what must be controlled.

Start with a strong sentence describing as completely as possible just what this policy will protect.

Who is responsible? Include all levels of responsibility: who is responsible for abiding by the policy and who is responsible for ensuring compliance.

When does the policy take effect? There are at least two dates to consider. The first date is for existing employees and the second is for employees who join the organization after the policy has been adopted.

Several other questions, although not usually part of the published policy, are generally included in a cover letter:

Where is the policy enforced? Is it only for a specific group of employees at company-provided facilities or does the policy apply outside the physical plant of the organization? For example, does the policy affect the employees when they travel, and does it apply domestically, internationally, or both?

Is policy enforcement different at organization headquarters than it is in remote locations? Is enforcement different in different areas? Do the accountants have a different set of controls than the research engineers? Or is the policy in effect everywhere?

Why has the policy been developed? This question leads into the sales aspect of your job. Your goal here is to explain why the controls should be established.

How will the policy be implemented? Will there be a gradual phase in or will it be effective on the publication date? How will the organization ensure that all employees have received adequate training or been made aware of the policy's existence? Furthermore, how will the employees be monitored?

When writing the policy, you should eliminate all unnecessary technical jargon. Each department within the organization will have its own language—

accounting, auditing, engineering, security, data processing, or any other department. To be effective, your policy has to be written in the common business language of your organization.

Terms like IMS, CICS, TSO, RACF, ACF2, bits, bytes, baud rates, modems, and so forth are so often used by people in data processing that they assume they're understood by everyone. But for an employee in accounting, those terms may be meaningless. To make sure you have succeeded in communicating to all the employees, someone outside your work area should review and criticize your work. The policy development department is the likely candidate for critiquing your work; it will make certain that your policy is in the proper format, and will identify jargon so it can be eliminated.

A policy statement should not exceed one or two pages. The wording must be concise so the reader can have no doubt about the objectives. The wording must also be unambiguous, so that no one can exempt himself or herself from the requirements.

Finally, know your readers. Knowing who will be reading you policy will have a direct effect on who it will apply to and how universal its acceptance will be. Perhaps you will need to modify the policy for different locations of the organization. If you work for a multinational organization, then you will have to change the policy statement to meet local requirements within the foreign locations. Try to be included in the review panel for any of these activities. Additionally, if your organization is in the defense business, then there are specific Department of Defense regulations that must be followed.

WRITTEN POLICY CHECKLIST

While the format of the policy will depend on your organization's publication style, specific items should be included, or at least considered, for each policy. The written policy normally includes, the following:

1. A declaration of what is to be protected or the purpose of the policy.
2. The audience for the policy.
3. The parties responsible for enforcement.
4. The ramifications of noncompliance.
5. The deviations, if any, that are allowed.
6. A description of the deviation process.
7. The audit trails that are available or required.

In addition to the preceding items, the employee community should be aware of the following:

- What is the author's job department or area?
- What department has update responsibility?
- How often will the policy be reviewed and updated?

• What is the date of the last revision?

EXAMPLES OF POLICY STATEMENTS

Having just learned what to include in a policy statement, this section will give you some examples of corporate policy statements. As you review the examples, use the checklist to see how many items each policy incorporates.

EXAMPLE I **UTILITY COMPANY**

Computer systems are company assets and must be protected from accidental or unauthorized disclosure, modification, or destruction. All employees and contractors have an obligation to protect computer resources by adhering to good security practices.

(Checklist items 1 and 2)

EXAMPLE 2 **INSURANCE COMPANY**

The protection of assets such as employees, physical property, and information relating to the conduct of business is a basic management responsibility. Managers are responsible for identifying and protecting all assets within their assigned area of management control. Managers are responsible for ensuring that all employees understand their obligation to protect company assets; they are responsible for implementing security practices that are consistent with generally accepted practice and with the value of the asset. Finally, managers are responsible for noting variance from established security practice and for initiating corrective action.

(Checklist items 1, 2, and 3)

EXAMPLE 3 **MAJOR U.S. BANK**

Information is a principle asset of the bank and must be protected to a degree appropriate to its vulnerability and its importance to the organization. Responsibility for information security rests with all employees on an ongoing basis.

(Checklist items 1 and 2)

EXAMPLE 4 **WASHINGTON-BASED BANK**

Information is a bank asset. Its access, use, or processing, whether on company-provided processing devices or noncompany-provided devices, at any site, is under the authority of all applicable regulations and corporate policies and standards. Steps shall be taken to protect information from unauthorized modification, destruction, or disclosure, whether accidental or intentional. It is the standard of the bank to limit access to information on a need-to-know basis to the smallest subset of authorized users.

(Checklist item 1)

EXAMPLE 5 **CALIFORNIA-BASED BANK**

Information is a bank asset, and, as such, steps shall be taken to protect it from unauthorized modification, destruction, or disclosure, whether accidental or intentional. The cost of such protection should be commensurate with the value of the information and the probability of the occurrence of a threat.

(Checklist item 1)

EXAMPLE 6 **MANUFACTURING CORPORATION**

Other than as specifically authorized by the appropriate company section manager, use of computer assets is restricted to company business. When using computer assets, each employee is authorized to access only that information that is required to do his or her job. Unauthorized access to any other information is strictly prohibited.

(Checklist item 5, and maybe 1 and 2)

EXAMPLE 7 **GLOBAL MANUFACTURING CORPORATION**

The protection of corporate assets, such as physical property and information relating to the conduct of business, is a basic management responsibility. Managers must identify and protect all assets within their assigned area of management control; they are responsible for ensuring that all employees understand their assigned area of management control. Managers are responsible for ensuring that all employees understand their obligation to protect company assets; they are responsible for implementing security practices that are consistent with those discussed in the Corporate Information Security Manual. Finally, managers are responsible for noting variances from established security procedures and for initiating the required corrective action.

(Checklist items, 1, 2, 3, 5, and 7)

EXAMPLE 8 **TELECOMMUNICATIONS COMPANY**

Purpose

The purpose of this policy is to help drive the company as a world-class leader in the internal use and management of information and data, to streamline the business processes and internal operations, and to enhance our competitive position in the marketplace. In this policy, the terms information and data are used based on the following definitions. Information is an aggregation of data that is used for decision making; data is the representation of discrete facts.

Policy
- Information and data are corporate resources.
- Data will be safeguarded.
- Data will be shared based on company policies.
- Data will be managed as a corporate resource.
- Corporate data will be identified and defined.
- Databases will be developed based on business needs.

- Information quality will be managed actively.
- Information will be utilized to enhance current offerings and pursue new business opportunities.

Goals

Successful management of information and data is critical to the operation of the company. Through active planning, organization, and control of these corporate resources, we will:

- Manage information as a strategic asset to improve the profitability and competitive advantage of both our customers and the company.
- Implement databases that are consistent, reliable, and accessible to meet all corporate requirements.
- Maximize the business processes through excellent data management across all business units.

Responsibilities

Every manager is responsible for implementing and ensuring compliance with the Company Information Policy and initiating corrective action if needed.

In implementing this policy, each business unit head is responsible for:

- Communicating this policy to employees.
- Establishing specific goals, objectives, and action plans to implement the information policy and monitor progress in implementing the policy.
- Developing plans that drive information systems and database development to satisfy both customers and corporate information needs.
- Actively supporting strong data management through data stewardship.
- Providing education and training in data management principles to employees.

The policy has been signed by the Chairman of the board.

(Checklist items 1, 2, 3, and 7)

CHAPTER REVIEW

1. Every data processing installation should have security procedures. Security requirements can be mandated by existing federal, state, or local laws, regulations, or contract.
2. Senior management is responsible for ensuring that policies and procedures are published and current.
3. Security procedures should be examined to ensure that they add to the organization's needs and do not inhibit business objectives.
4. The written policy should prevent confusion, not generate it.
5. Unnecessary technical jargon should be eliminated from the policy statement, and the wording should be unambiguous.

Exercise 1

Using the information contained in Chapter 1, develop a policy statement for your organization.

• Remember the checklist items.

• Allow for any legal and contractual agreements or regulations.

• Use the examples given as guidelines for your policy.

Exercise 2

Read the following generic policy statement:

It is the policy of this company that all information utilized in the course of business is considered an asset, and, as such, managers and employees are responsible and accountable for its protection. It is a management responsibility to maintain information security and integrity through the development and administration of appropriate controls to protect company information from intentional or accidental disclosure, modification, destruction, or denial of access.

What, if anything, is missing or unclear about this policy?

Would an appropriate usage and responsibility statement increase the overall effectiveness of this policy?

CHAPTER NOTES

1. Courtney, Robert H., "Management, Auditing, Security—Changing Responsibilities," *Datamation*, September, 1985.

2. U.S. Congress, Senate Committee on the Judiciary. *Computer Fraud and Abuse Act of 1986*. Proceedings of August 16, 1986.

3. Horton, Thomas R., "The Business of Managing Information . . . Securely from One CEO to Another," *Datamation*, September, 1989.

4. Groves, Ray J., "The Oversight Roles in Protecting Information Assets," *Datamation*, September, 1987.

5. *Ibid.*

6. Bigelow, Robert P., "Top Management Needs to Know about Data Security and the Law," *Datamation*, September, 1984.

7. Horner, Larry D., "What Every CEO Needs to Know—and Do—about Computer Security," *Datamation*, September, 1986.

8. McGovern, Patrick J., "Planning for Information Security: Grassroots and Global Concerns," *Datamation*, September, 1988.

CHAPTER 2

Mission Statement

SETTING THE SCOPE OF THE MANUAL

The policy statement that you just created will establish your organization's position on the protection of assets, whether physical or intellectual. Your next goal is to write a mission statement that will define the areas that need to be protected.

Implementing the controls required to meet the objectives of your policy statement, no matter how large or small, can be an overwhelming task. A well-written and properly-endorsed mission statement or charter will allow you to focus on the areas that require control. The mission statement will also educate employees about the overall direction of your assignment. By stating your mission, you will be laying the groundwork for the success, acceptance, and incorporation of the policies and procedures you have created.

Most organizations look on mission statements or charters as enabling acts. That is, they establish the scope of responsibility for each department or individual. Most organizations have a procedures and methods division within the financial staff. The job of the procedures and methods division is to generate the accounting practices and procedures that will ensure that your organization complies with the generally accepted accounting practices (GAAP) standard. Unfortunately, the field of information and computer security has no equivalent. Therefore, each organization must start from scratch when attempting to develop computer and information security policies and procedures.

BACKGROUND ON YOUR POSITION

Before you can begin working on your mission statement, you will have to understand two important issues. The first issue concerns why you are here. This isn't an open-ended question about your existence, but rather why are you writing computer or information security policies and procedures? Most people would rather be doing something else, something exciting and visible, anything but writing procedures. The second issue is why management assigned the task in the first place. Management does not undertake a policy review unless it is getting pressure from another source, such as senior management or open audit comments.

Before you begin to write procedures, you must first fully understand why you were selected to do the job and why management decided to develop procedures at this point in the life of the organization. The answers you receive will allow you to create a mission statement with an appropriate focus.

To create an efficient and effective set of computer security policies and procedures, it is best to begin the process at the corporate headquarters or main office and then solicit input from the divisions, groups, units, and subsidiaries and from certain identified individuals. By involving as many of the business units as possible, the level of resistance may be reduced.

Because of varying organizational methods (and because you want to ensure your job won't be eliminated), you must understand fully the organizational structure and the events that led to assigning you to the project. Once you have this information, you can create an outline of specific goals that will be used to meet the policy objectives. When the outline is completed, you will want senior management to approve the development of your mission statement. Only when the mission statement is completed will you be in a position to begin writing standards and procedures.

The section on policy development in Chapter 1 should have given you an understanding of the global impact and need for protecting information and computer systems. In addition, Chapter 1 gave you the necessary background for writing a policy statement, and you will use this information over and over again because each section of your procedure manual will require a policy or overview statement. By understanding the global impact of your security guidelines, you will be better able to enlist the support of management. Your current assignment, however, may not be as sweeping and grandiose as establishing organization-wide policies and procedures. By learning why you have been enlisted to write procedures, you will have a good perspective for generating your mission statement.

BUSINESS GOALS VERSUS SECURITY GOALS

People who work in computer departments often focus on their immediate concerns and in the process lose sight of the organization's goals. Every organization has an overall business objective. To make your computer security program constructive, you must seek out the organization's business objectives and ensure that your security program meets and supports those goals.

As an employee in the department charged with the role of defining security controls, you must acknowledge that security is not your organization's most important product. Any computer and information security objectives you develop must be adapted to meet the practical business conditions of your organization. Security controls that inhibit the business function of the organization will be quickly discarded. Poorly written procedures will not support management's fiduciary responsibility to protect assets.

Because of the inherent difference between protecting information and promoting business, your job assignment may often appear to be in direct conflict with the objectives of the rest of the organization.

After all, the most secure computer system is one that is down. However, although the data is safe from unauthorized access, having the system down does not meet any business goals or justify its installation in the first place. So your goal is to ensure that when the computers are operational, only those employees with a legitimate need for access will be granted access, and they will be allowed access on a need-to-know basis. With proper controls in place, the organization will have maximum security with a minimum of impact, and the security will be cost effective.

Your organization's business objectives are usually stated in readily available publications, such as:

• Annual reports to stockholders
• Organizational charts
• Strategic planning information
• Interviews with staff members
• Annual corporate budget proposals

The annual report is especially valuable to you as a security officer. This document contains the management report, in which the chief operating officer and chief financial officer attest that there are adequate controls in place to protect the organization from loss of assets and prevent the organization from being at risk. By using this information, you will learn the organization's business objectives and gain senior management's support in the development of adequate controls.

COMPUTER SECURITY OBJECTIVES

Before reviewing existing mission statements and then creating your own, you must explore the elements of a comprehensive information security program. Remember, computer security is just one part of the organization's overall asset protection program. While you will address the physical security of computers, and you must protect the considerable investment in computer hardware, you must realize that the computer only functions as the processor and storage device for information. The information, data, programs, applications, transactions, and systems are extremely valuable, and they must be protected as well as the hardware. If the data has been backed up, then the system can eventually be brought back up. However, if the data has not been backed up, nothing can bring the system back. So be sure to include controls on the assets as well as the hardware.

The following items are generally accepted standards for a comprehensive security program:

- Ensure the accuracy and integrity of data.
- Protect classified data.
- Protect against unauthorized access, modification, disclosure, or destruction of data.
- Ensure the ability of the organization to survive the loss of computing capacity (disaster recovery planning).
- Prevent employees from probing the security controls as they perform their assigned tasks.
- Ensure management support for the development and implementation of security policies and procedures.
- Protect management from charges of imprudence in the event of any compromise of the organization's information or computer security controls.
- Protect against errors and omissions (which still account for 75 to 80 percent of losses).

MISSION STATEMENT FORMAT

Most mission statements begin with a brief paragraph explaining the overall goals of the information and/or computer security program. This initial statement takes the concepts established in the policy statement and expands them into the goals to be addressed by your department. Your mission statement or charter should also reflect your organization's style. As discussed, you need the background on management's impetus for establishing policies and procedures and on the overall business objectives to begin writing a mission statement.

After the section on goals, the mission statement should then list your responsibilities. The responsibilities are usually presented in active voice and provide the key elements of your job description. The effectiveness of the security program you develop and your effectiveness as an individual employee will be judged on the basis of the responsibilities you describe.

MISSION STATEMENT EXAMPLES

EXAMPLE I **GLOBAL MANUFACTURING CORPORATION**

As operations and related information processing become more decentralized through the use of PCs, controls and procedures become more important. And as remote access, networks, and distributed information processing make possible still greater globalization and decentralization of computer-generated data, the potential for unauthorized access to company secret, confidential, and restricted information increases. To provide the corporation with the highest level of visibility and support for the philosophy of information security and to provide the groups, units, divisions, sections, staffs, and departments with a focal point for solving security problems, a Corporate Information Security Group has been established and will report to the Director of Security.

Corporate Information Security Group Responsibilities:

- Keep information and computer security policies and procedures current.
- Answer all inquiries on compliance and interpretation of corporate policies.
- Review all computer and information audit comments and the associated responses, thereby providing independent review of audit comments and unit responses.
- Review the employee Security Awareness Program to ensure that it remains an effective tool for information security controls.
- Assist departments in developing recovery plans and oversee the testing of these plans.
- Review new computer and information security products and make recommendations on these products to ensure they meet minimum corporate requirements.
- Assist in the investigation and reporting of computer equipment thefts, intrusions, viruses, and breaches of information security.
- Assist local Information Security Officers in developing effective training programs for the Plant Security Officers in the areas of computer crime and investigation.
- Assist in the development of effective monitoring programs to ensure that corporate information is protected as required.
- Ensure that the systems moved into production mode are safe from errors and omissions.

The preceding statement has a broad scope to accommodate all the sites, both foreign and domestic, of a large organization. While this is an effective charter for a multinational corporation, it may not suit the needs of your company. The next examples narrow the scope of responsibility and may be more in line with your needs.

EXAMPLE 2 **NORTH AMERICAN MANUFACTURING COMPANY**

An Information Security Administration Function (ISAF) shall be developed to establish standards, procedures, and guidelines as deemed necessary to ensure the security of information throughout the company. The "owners" of information (as defined later) will be required to take prudent security measures, with the assistance of the ISAF, to protect information from unauthorized modification, destruction, or disclosure, whether accidental or intentional.

The role of the ISAF is to develop written documents to ensure the security of company information. Beyond that, the specific duties of the ISAF are vague. This mission statement lacks items that can be completed by the ISAF.

A brief note about the term security: Security is one of those words that everyone interprets differently. In this book, security is synonymous with protecting company assets, both physical and intellectual. Protection is an active process that motivates employees to safeguard the assets of the organization.

EXAMPLE 3 CORPORATE DATA PROCESSING DEPARTMENT

The Information Systems Security Officer (ISSO) has been established to ensure that corporate data is protected from unauthorized modification, disclosure, and destruction and ensure that the corporation has security measures to carry out its responsibilities as defined by law and the courts.

Data Security Goals

- Develop a uniform protection policy.
- Have a data classification system to aid business managers in evaluating their data assets.
- Identify owners for all data sets.
- Have operators of the off-site data storage facility function as data custodians.
- Install and administer an access control package with a minimal use of passwords.
- Educate all users on the importance and use of security measures.

While the preceding goals are certainly admirable, there are no corresponding statements that define how the ISSO will implement these goals. More concrete recommendations are necessary for developing this company's security program.

EXAMPLE 4 CORPORATE INFORMATION SECURITY ADMINISTRATION

Contribute to the Corporate Information Security Program by performing the following tasks:
- Implement and coordinate the Corporate Information Security Policy and Program.
- Design and implement a Corporate Security Awareness Program.
- Design a strategy for detecting actual security risks.
- Prepare company-wide policies.

Assist management in performing their security responsibilities by performing the following tasks:
- Assess proposed access controls.
- Prepare and publish guidelines and standards.
- Assist areas in the development and enforcement of internal security procedures.

- Ensure that criteria for sensitive and critical information are current and appropriate to the needs of the corporation.
- Train the area coordinators in maintaining and enforcing guidelines and standards.
- Participate in the application development cycle.
- Advise on contingency planning and disaster recovery plans.

Prepare and monitor management processes to prevent and handle perceived information access violations by performing the following tasks:

- Perform reviews of all information security access control systems.
- Ensure that appropriate information security requirements are being enforced.
- Approve all events in which established information safeguards are overridden and ensure that each override is documented.
- Ensure that security violations are reported to the appropriate manager.
- Contribute to the annual audit report on information security and access.

Recommend allocation of resources and technology enhancements to meet information security objectives by performing the following tasks:

- Select and administer all information security access control systems.
- Review existing and proposed hardware and software for security considerations and make recommendations, as appropriate.
- Delegate limited administrative authority to other individuals or groups, if appropriate.
- Execute a risk assessment of sensitive data and the cost of protecting it.

This is an aggressive mission statement, but it may also be unrealistic. The CISA will assume the responsibility for security measures throughout the corporation. Unless there is adequate staff, the objectives are far too broad for a typical computer security department to undertake. If you develop a detailed mission statement like this one, make certain that you are not the only person assigned to work on computer security.

EXAMPLE 5 MEDIUM-SIZED MANUFACTURING COMPANY

Introduction

This document defines the scope and direction for the information security function. The duties and responsibilities set forth will serve as the charter for the group.

Responsibilities of Management

To fulfill present and future business commitments, steps must be taken to ensure the accuracy, privacy, and security of our computers, communication networks, electronically processed data, and manual data. The responsibility for safeguarding corporate information rests with all employees, but it is the coordinated effort of management and information security that will:

1. Minimize the probability of security breaches.
2. Minimize the damage if such a breach occurs.

3. Ensure the company's ability to recover from damage with minimal disruption of service.

It is a basic management responsibility to protect resources necessary to conduct business. Management is responsible for identifying and protecting hardware, software, and data resources under its control. This task is accomplished by implementing security policies and practicing security procedures commensurate with the value of the asset to the company.

Responsibilities of Information Security Management

Mission

To provide a secure environment for the information assets of the company.

Strategies

Monitor and audit adherence to security policies and procedures on a daily basis.

Maintain an ongoing and corporate-wide security awareness program relating to information asset protection.

Act as a catalyst to make security a part of each employee's daily activities.

Ensure that the company has adequate protection for its business information assets and the most cost effective tools to eliminate security breaches.

Maintain an ongoing security audit process to review security exposures or breaches in a timely manner.

Key Responsibilities

Establish and enforce the following general data security rules in conjunction with management:

- Information shall be created and maintained in a secure environment.
- Practices shall be in place to prevent unauthorized modification, destruction, or disclosure of information, whether accidental or intentional.
- Safeguards shall be implemented to ensure the integrity and accuracy of vital company information.
- The cost of information security shall be commensurate with the value of the information to the company, the company's customers, and potential intruders.

Formulate an overall security plan for the corporation.

Review company information security practices regularly, considering technological, environmental, and statutory requirements and trends. Keep abreast of new security developments that could affect the company.

Perform reviews and/or act as a consultant in matters affecting information security.

Provide support to all employees as they fulfill security-related responsibilities.

Perform security administrator duties in areas where direct responsibility for information security has been assigned.

Conduct periodic risk analysis inspections of data processing facilities and software systems to identify security exposures and report the findings to the respective management.

Develop, maintain, and implement policies, procedures, and guidelines to assure information security.

Assist plant security to develop, maintain, and implement policies, procedures, and guidelines to assure the physical protection of information assets.

Provide information security awareness training to all company personnel.

Coordinate the installation and maintenance of security software on systems for which direct responsibility has been assigned. Monitor the installation and maintenance of security software on all company hardware that is under the control of remote site security administrators.

While this mission statement is longer than is typical, it is thorough. Management responsibilities are defined from the start, and these set the tone for what is expected from each segment of the company—management, employees, and security. Nevertheless, the section on strategies does not need to be shared by the company as a whole and probably should be an internal departmental function.

EXAMPLE 6

Charter

The mission of the Information Security Department (ISD) is to direct and support the company and affiliated organizations in the protection of their information assets from intentional or unintentional disclosure, modification, destruction, or denial through the implementation of appropriate information security and business resumption planning policies, procedures, and guidelines.

Responsibilities

The ISD shall be responsible for the development and administration of information security control plans, including the following tasks:

1. Develop information security policies, procedures, and guidelines in compliance with established company policies and generally accepted data processing controls.
2. Implement a data classification system and a management assessment program to be completed annually.
3. Develop and maintain a company-wide information security awareness and education program.
4. Develop and maintain an overall access control program for mainframes, minicomputers, and microcomputers.
5. Select, implement, test, and maintain an appropriate business resumption plan for each company location responsible for processing critical systems and applications.
6. Ensure that information security requirements are incorporated in new applications by participating in the systems design and development process.
7. Investigate and evaluate emerging information security technologies and

services and coordinate implementation of appropriate hardware, software, and services within company operating groups.

8. Coordinate the distribution of company security information and provide technical assistance to operating organizations as required.

9. Implement other information security responsibilities as deemed appropriate.

This mission statement is strong, with the exception of item 5. Taking on the role of facilitator for the company's business resumption plan should not be an add-on responsibility for the ISD. Business resumption planning (BRP) is a full-time job. In fact, BRP is an entire industry separate from computer and information security activities. Your role in a company BRP should only be the part that relates directly to data processing. A more appropriate responsibility statement would be: "Assist the departments and other business units in developing local business resumption plans and act as observer while these plans are being tested." For more information on contingency planning and how it affects computer security programs, see Chapter 4.

Your mission statement should spell out the goals that you believe can be accomplished. You have the opportunity to determine the direction your job will take, so be sure the responsibilities are attainable. Do not write a mission statement that assumes that you will have a staff of security personnel at your disposal. In the real world, the information security function is often a one- or two-employee operation, and security is just one of the assigned responsibilities. Be realistic about what you can accomplish and be certain to include educational responsibilities. For example, an appropriate item might state: "Attend workshops, seminars, and conferences annually to remain current on new developments in security technology."

SUPPORT FOR THE MISSION STATEMENT

Before publication, the mission statement must receive management approval. While the format of the mission statement does not really affect how it will be received, it is extremely important to have the statement approved at the highest possible level of management.

The following examples show typical approval levels of mission statements for established organizations.

- General Motors Corporation—Chairman
- KMart Corporation—Chairman of the Executive and Finance Committee
- Capital Holding Corporation—Chairman of the Board and Chief Financial Officer
- AT&T New Jersey—Chairman of the Board
- Miller Freeman Publications—Member, Board of Directors

It is important to note that in the preceding examples no data processing personnel approved the mission statement when published. Although an effective security program can be established and flourish with only data processing approval, the overall acceptance and support will be greatly reduced and your task will take a much longer time to succeed.

KEY ROLES IN ORGANIZATIONS

This section describes some of the different roles within an organization and the responsibilities associated with each job. This section will provide the groundwork for identifying the management levels within your organization. The specific roles within your organization may vary, but you can find corresponding positions in your organization for each of the following key management functions.

Chief Executive Officer (CEO). A member of an organization who has authority over all other members in determining the conduct and direction the organization will take. The CEO is elected as a director by the shareholders and appointed CEO by the board. The CEO is responsible to the shareholders of the company for the successful conduct of the company.

Chief Financial Officer (CFO). Along with the CEO, the CFO is responsible for maintaining a system of internal controls designed to provide reasonable assurance that the books and records reflect the transactions of the organization and that its established policies and procedures are carefully followed. Perhaps the most important feature in the system of control is that it is continually reviewed for effectiveness and is augmented by written policies and guidelines, the careful selection and training of qualified personnel, and a strong program of internal control.

The CEO and CFO must sign an annual consolidated financial statement attesting that the organization has adequate controls to protect vital assets. The audit staff reports provide the information for this critical function.

EXAMPLE **ANNUAL REPORT**

Management Report

Company management is responsible for the fair presentation and consistency of all financial data included in this Annual Report in accordance with generally accepted accounting principles. Where necessary, the data reflect management's best estimates and judgments.

Management also is responsible for maintaining a system of internal accounting controls with the objectives of providing reasonable assurance that Company assets are safeguarded against material loss from unauthorized use or disposition and that authorized transactions are properly recorded to permit the preparation of accurate financial data. Cost-benefit judgments are an important consideration in this regard. The effectiveness of internal controls is maintained by: (1) personnel selection and training; (2) division of responsibilities; (3) establishment and communication of policies; and (4) ongoing internal review programs and audits. Management believes that Company system of internal controls as of December 31, 1989, is effective and adequate to accomplish the above described objectives.

(signed) Chairman and Chief Executive Officer
(signed) Senior Vice President and Chief Financial Officer
February 23, 1990
Source: Monsanto 1989 Annual Report, page 20.

Senior Management. The senior manager of a business unit, such as the director of accounts payable, is responsible for specifying and implementing the operational controls for his or her work area. In addition, this individual is considered the owner of the information assets for the department he or she oversees. Senior management is responsible for ensuring that controls are in place to safeguard the department's data, including who may read and update files. Senior management may delegate the responsibility for the day-to-day approval process for access to the data, programs, and transactions to another employee within the department. However, because the senior manager and the delegate are responsible for the routine reconciliation of the department's activities, their positions should not include the ability to originate data or transactions. (The separation of duties is discussed further in the chapter on data classification.)

Director of Management Information Systems (DMIS). The DMIS is the highest level of management within the organization charged with responsibility for the operation of the computer systems (not including microcomputers). This individual is responsible for ensuring that systems programmers, application programmers, system operators, and scheduling, tape library, and other related personnel are conducting their daily activities in accordance with established policies and procedures. The DMIS is also responsible for the actions of the system security administrator and any privileged users.

Information Security Officer (ISO). The ISO is responsible for developing the computer and information security policy to be adopted by senior management. The ISO is also responsible for advising on protective measures (including standards and procedures), measuring performance, and reporting to management. The ISO may supervise the system security administrator(s).

System Security Administrator (SSA). The SSA is responsible for creating and maintaining access control records. The SSA acts as surrogate for the system manager and the application and data owners. The SSA enrolls new users

FIGURE 2 **DATA PROCESSING ORGANIZATION CHART**

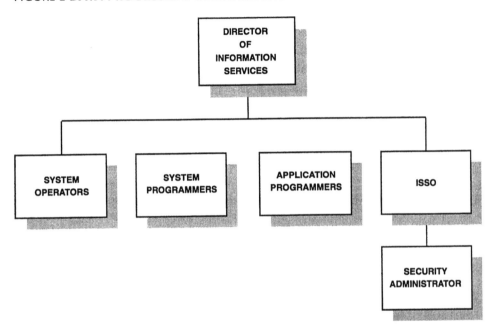

and grants access. The SSA works under the supervision of the director of information services or the ISO and is subject to review by internal auditors.

If you are not clear where you fit into your corporate organization, you should obtain some current organization charts and find out who your ultimate boss is. Figure 2 shows one example of a data processing organization chart. In most instances, data processing falls under the responsibility of the financial staff. However, recently in companies such as IBM, Electronic Data Systems, BP America, Shell, and Aetna, the information security officer reports to the head of security. Knowing where you report in an organization will help you develop a mission statement that will support the business goals of your organization.

BUSINESS OBJECTIVES

Most computer security policies and procedures are based on the recognition that the organization is inextricably linked to computer systems. Without automated information processing, the corporate world would be unable to design new products, manufacture existing products, sell the products, or even collect money for the product or services rendered. Without the ability to access information in a timely and efficient manner, businesses would cease to function within days or even hours. As a result, management must understand that corporate information is vulnerable to errors, omissions, and unauthorized access, as well as modification, disclosure, and destruction.

As a member of the corporate team, computer security must present its goals and objectives in the format and language of the organization. In addition to the policy statement and mission statement, a five-year plan should be developed. Most organizations establish such plans to determine the overall direction. Like other departments of the corporation, computer security should establish its own short-term and long-term objectives. Once developed, the information security five-year plan should be reviewed annually and modified as necessary. During the annual review, you can list goals that have been completed, determine the status of ongoing projects, and prepare new updated, long-term objectives.

The business plan should support the goals established in the mission statement. Start with short-term goals that you are fairly sure you will be able to complete. Remember, nothing succeeds like success. When management sees that you are accomplishing your stated objectives, support for the security program will be easier to obtain.

CHAPTER REVIEW

The mission statement should ensure that the security of the information and communication processing resources of the corporation are sufficient to reduce risk to a level acceptable to the management of the corporation.

Responsibilities

- To recommend policies, standards, and procedures that foster the protection of information and information processing resources.
- To assist units and divisions in the selection and implementation of the protective measures required in their areas responsibility.
- To evaluate new technology recommend security strategies to protect it.
- To identify areas of potential risk in the protection of corporate computer and information assets and to alert management once those areas have been identified.
- Provide training for security control requirements during all phases of application and system development.
- Develop programs to increase security awareness at all levels of the corporation.
- Develop a liaison between the corporate security and audit staffs to ensure that security efforts are coordinated and resources are conserved by preventing duplication of effort.
- Coordinate and assist in the development of business resumption plans for all data centers supporting critical business functions.

- To work with the local ISSO to ensure that corporate-mandated programs are cost and operationally effective.
- To act as a consultant to all areas on the security of information and computer systems.
- Monitor changes in laws and regulations as well as changes in technology and corporate goals to determine the impact of these changes on corporate security requirements.

Exercise 1

Using the information in this chapter, develop a mission statement for your specific job description.

Remember to include an overall mission statement and to delegate areas of responsibility.

Exercise 2

Using the following organization chart, fill in the blanks for your own organization. Start with the chief executive officer and fill in the chart to your level.

FIGURE 3 **BLANK ORGANIZATION CHART**

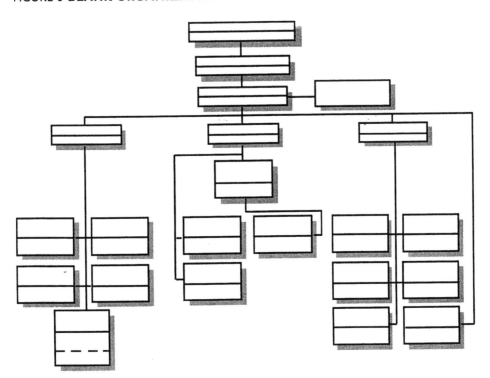

Exercise 3

Based on the organization chart completed in exercise #2, who should approve your mission statement? Who will probably sign your mission statement? Are these two different? Why?

Exercise 4

As a member of the corporate team, complete an initial five-year plan for the computer or information security department. When preparing your plan, be sure to answer the following questions.

- What are the objectives and how do they support the business goals of the organization?
- What are the most pressing issues, and what needs immediate attention?
- Which problems can be corrected with the least amount of capital outlay?

CHAPTER 3

Data Classification

Classification establishes a system of priority for protecting corporate resources. With an effective classification system, the organization can ensure that protection levels are commensurate with the value of the information or system being protected. By classifying levels of information, the organization is able to reduce the cost of protection, since not all information requires the same security.

While overprotection is always possible, problems typically come from underclassifying information and from allowing too many employees to have access to it. Recent audit findings and reports from local computer security special interest groups' meetings indicate that unnecessary access continues to be a major security problem.

Information classification allows an organization to protect the value and limit the exposure of the resource. With written security guidelines, information owners are responsible for classifying the information and protecting it, thereby minimizing exposure of the asset. The classification process ensures that the controls are enforceable and that costs will be kept to a minimum.

LEGAL BASIS

In the United States, a number of laws require organizations to establish an effective information classification system. Due to changes in legal statutes, you should check with your legal staff to learn the current status of computer and information laws in your locale. Additionally, Chapter 9 provides a brief discussion of all current U.S. statutes and a summary of some foreign regulations.

CLASSIFICATION REQUIREMENTS

Classified data is information developed by the organization with some effort and some expense or investment that provides the organization with a competitive advantage in its relevant industry and that the organization wishes to protect from disclosure.

While defining information protection is a difficult task, four elements serve as the basis for a classification scheme:

1. The information must be of some value to the organization and its competitors so that it provides some demonstrable competitive advantage.
2. The information must be the result of some minimal expense or investment by the organization.
3. The information is somewhat unique in that it is not generally known in the industry or to the public or may not be readily ascertained.
4. The information must be maintained as a relative secret, both within and outside of the organization, with reasonable precautions against disclosure of the information. Access to such information could only result from disregarding established standards or from using illegal means.

INFORMATION CLASSIFICATION CLASSES

Company Secret

It is recommended that senior management first classify company secret information, using the following guidelines.

Company Secret Characteristics

1. The information provides the organization with a significant competitive edge.
2. Unauthorized disclosure of the information would cause substantial damage to the organization.
3. The information shows specific business strategies and organizational directions.
4. The information is essential to the technical or financial success of a specific product or service.

Examples of Company Secret Information

- Specific operating plans, marketing plans, or strategies.
- Descriptions of unique parts or materials, technology intent statements, or new technologies and research
- Specific business strategies and directions.
- Design and/or specification files.

Company Confidential

Company confidential information should be classified after company secret, since company confidential can have the same characteristics as company secret. The distinction between the two depends on management's evaluation of the importance of the information (value, competitive advantage), the number of personnel granted access to the information, and the amount of effort necessary to protect the information. The differences between access and effort for company confidential and company secret are highlighted in the data classification matrices, shown in Figures 5 and 6 at the end of the section on classification.

Examples of Company Confidential Information

- Consolidated revenue, cost, profit, or other financial results.
- Operating or marketing plans and marketing strategies.
- Descriptions or unique parts or materials, technology intent statements, or new technological studies and research.
- Marketing requirements, technologies, product plans, or revenues.
- Personnel and payroll information.
- Credit or customer financial information.

Company Restricted

Company restricted classification indicates that unauthorized disclosure or use of information would not be in the best interest of the organization, its stockholders, its employees, or its customers.

Examples of Company Restricted Information

- Summaries of financial data for the organization's operations of position, such as investments, engineering expenditures, and manufacturing costs.
- Revenue, cost, profit, or other financial results for one of the organization's divisions or subsidiaries.
- Disclosure of the overall product or feature structure and unannounced product designs.
- Major changes in the organization's management structure.
- Personnel information including appraisal, career path, date of birth, and salary review.

Company secret, company confidential, or company restricted classification should not be confused with the Department of Defense (DOD) data classification. For additional information on DOD requirements, refer to the *Industrial Security Manual for Safeguarding Classified Information.*

Proprietary

Proprietary information is all business-related information requiring baseline security protection, but failing to meet the criteria for higher classification levels.

Examples of Proprietary Information

- Business data, such as organization telephone directories.
- Organization policies, standards, or procedures.
- Organization announcements (internal).

- Day-to-day reports, computer runs and general financial data of medium sensitivity.

OVERPROTECTION

When protecting data, the employees, and especially management, will have to take a close look at the types of information they handle, and, using the preceding guidelines and the matrices in Figures 5 and 6, determine the degree of protection required for each type of data.

Classification applied to information should be kept as low as possible without compromising the security of the data. Obviously, some organizations, such as military sites, cannot tolerate unauthorized access to sensitive information, but most organizations do not need to enforce such rigid standards. Overclassification comes at a cost and may slow down your system because of the extra precautions required for secure handling and storage. If almost all data is company secret or company confidential, then there actually is no classification scheme, and enforcement will be difficult. Finally, information that is overclassified will soon cause the employees to disregard the classification system, and the classification of confidential information will lose its value.

When information does require protection, make sure that the protection is consistent. Often strict access controls are applied to data stored in mainframe computers but are not applied to other office hardware. Whether in a mainframe, minicomputer, microcomputer, file cabinet, desk drawer, waste basket, or in the mail, information should be subject to consistent rules.

One general security principle is that classified material should be locked away when not in use, particularly after office hours. The simplest way to ensure this is to adopt a "clean desk" policy so that all papers are put away at the end of the day. This regulation has, as you may guess, been a major sticking point for organizations that try to enforce it. One compromise is to create high security areas (HSA). HSAs are established within local units to control access to company secret information. Access to HSAs is restricted to company employees with an authorized access need. Outside the HSAs, the "clean desk" policy can be encouraged. However, within HSAs containing company secret information, a "clean desk" policy should be required. Managers responsible for information security should make periodic spot checks after office hours to determine the effectiveness of security procedures. Your overall information protection program should also include assigning security staff to make similar spot checks on a regular basis.

Unclassified

Unclassified information is information that is not sensitive in context or content and requires no security protection.

Examples of Unclassified Information

- Online public information.
- Internal correspondence, memoranda, and documentation that do not merit a security classification.
- Public organization announcements.
- Outdated information.

If the information does not fit these descriptions, it must be classified and protected as discussed in the preceding sections.

Other Classifications

The preceding classification levels are routinely used in business environments. There are also many other acceptable classification labels in use, and these labels may be just as effective for your organization. Some of these classifications are:

Critical. Also used in the development of disaster recovery planning. Information or systems considered to be integral to the business and without which operations would be curtailed or otherwise severely impeded.

Essential. Information or systems less essential to the business than information or systems classified as critical and without which operations would be difficult but not severely curtailed or impeded.

Internal-Use-Only. Information to be used by organization employees only, similar in scope and control to information in the company restricted category.

Personnel/Private. Information regarding the employees. By law, physical, technical, and administrative safeguards must be implemented to ensure the security and confidentiality of the data. Normally this information is the responsibility of the personnel staff, but payroll and medical records are also included.

Privileged. Information concerning a lawyer-client relationship.

Public. Information that is not classified and is on public record, requiring no level of protection.

Sensitive. Information holding any classification level. Information classified as secret, confidential, restricted, and proprietary are all Sensitive classification levels.

Vital. Data elements in critical applications and systems that require appropriate backup and off-site storage.

No matter what the classification scheme, adequate control must be maintained over your organization's classified information. This information, whether in electronic or hard copy format, must be safeguarded while it is being prepared, mailed, used, stored, or purged. When copies are required of confidential information, only a minimal number should be made, and employees must be made aware of their responsibility to safeguard the copies as well as the originals. The procedures you develop must determine whether or not copying of company secret information should be allowed.

Declassification

The sensitivity of most classified material decreases over time. The document originator may determine the date when protection will end, for example, 'company confidential until (date)'. Declassification can also be tied to a public statement, as in the quarterly earnings reports. Prior to the announcement of quarterly earnings, the information is classified as company confidential. Once the announcement is made, however, the information reverts to the unclassified category.

Destruction

Part of an effective records management program is to destroy documents when they are no longer required. Placing restrictions on copying classified documents will ensure that logs are created to audit the number of copies and the users for whom the copies are made. In a typical security program, the originals would be sent to a records management center and all copies would be destroyed by the recipient. Verification that the document had been destroyed would be sent to the originator.

For an overview on the issue of records management, you should consider contacting Commonwealth Films in Boston to obtain a copy of the video *Buried Alive*. Further information on computer security films that are currently available is provided in the Appendix.

Information Labeling

The debate over labeling and labeling data has raged for years. The pro-labeling group believes that when a document is properly labeled with its classification level, the recipient will know how to handle the document. The anti-labeling group feels that labeling a document as classified calls more attention to it and encourages probing by unauthorized users. If you do not label the material, however, you will not have followed the fourth element for classifying information, and you may have trouble defending your security system in a court of law. As a result, companies are now moving toward labeling documents that meet the security guidelines. By labeling classified information, the organization establishes a clear set of standards for the treatment of the material. Employees will not be able to plead ignorance as an excuse for mishandling sensitive data.

Proper classification by physical marking, notation, or other means informs and warns the user about the degree of protection required for that information. All highly classified information and material must be marked in such a manner that the assigned classification level can be clearly ascertained.

All company secret and confidential material should be marked as follows:

1. The document should contain the name and address of the facility responsible for its preparation and the date of preparation.
2. The document or any reproduction of it should be stamped or marked company secret or confidential at the top and bottom of the outside cover (if applicable) or on the title page.

Employee Instructions

While an open climate of information sharing is desirable to satisfy the needs of the business and its employees, there is a clear need to safeguard organizational information. Access to classified organizational information should be based on a clear business "need to know." Classified organizational information should not be discussed with family or friends, as such discussions may lead to disclosure, and disclosure may be detrimental to the business interest of the organization.

Access to information does not imply or confer authority to act as spokesperson for the organization concerning such information or to discuss such information with others. Public announcement of classified information relieves an employee of his or her responsibility to maintain security only to the extent of the information included in the announcement. Speculative press reports provide no excuse for comment on or disclosure of organizational information. Substantial competitive advantage may be sacrificed through untimely disclosures and may result in a loss of customer confidence and business.

FIGURE 4 **DATA CLASSIFICATION MATRIX**

	SECRET	CONFIDENTIAL	RESTRICTED	PROPRIETARY	UNCLASSIFIED
Storage on fixed media	Encrypted	Unencrypted	Unencrypted	Unencrypted	Unencrypted
Storage on exchangeable media	Encrypted	Encryption optional	Encryption optional	Unencrypted	Unencrypted
Access to object	Data owner to explicitly define individual users	Data owner to explicitly define individual users	Data owner to define permissions	Available on a "need to know" basis	No restrictions
Initiate electronic copy of object	Data owner authorization required; Equivalent security clearance required	Data owner authorization required; Equivalent security clearance required	Equivalent security clearance required	No special restrictions	No restrictions
Initiate hard copy of object	Forced collaboration; Predefined manned device	Predefined manned device	Manned device	No special restrictions	No restrictions
Modification of object	Equivalent security clearance to data owner required	Equivalent security clearance to data owner required	Equivalent security clearance to data owner required	No special restrictions	No restrictions
Deletion of object	Must be data owner; System confirmation required	Data owner authorization required; System confirmation required	Data owner authorization required; System confirmation required	System confirmation required	No restrictions
Labeling of object	Display SECRET on computer screen	Display CONFIDENTIAL on computer screen	Display RESTRICTED on computer screen	No requirements	No requirements
Transmission over WAN	Encryption required; Electronic confirmation required	Encryption optional; Electronic confirmation required	Encryption optional; Electronic confirmation required	Unencrypted	Unencrypted
Transmission over LAN	Encrypt unless secured area; Electronic confirmation required	Encryption optional; Electronic confirmation required	Encryption optional; Electronic confirmation required	Unencrypted	Unencrypted
Labeling of computer media	Media must be marked SECRET	Media must be marked CONFIDENTIAL	Media must be marked RESTRICTED	Marking done at data owner's discretion	No requirements
Destruction of computer media	Physical destruction by "secure" method	Physical destruction by "secure" method	Physical destruction by "safe" method	Physical destruction by "safe" method	No special requirements
Audit	Log all access attempts; Subject to 100% management check; Retain for 5 years	Log all access violations; Subject to 100% management check; Retain for 5 years	Log all access violations; Subject to management check; Retain for 5 years	Log all access violations; Subject to management check; Retain for at least 1 year	No special requirements

FIGURE 5 **DATA CLASSIFICATION MATRIX: PRINTED DATA**

	SECRET	**CONFIDENTIAL**	**RESTRICTED**	**PROPRIETARY**	**UNCLASSIFIED**
Labeling of documents	Must be marked SECRET, registered with serial number, pages numbered	Must be marked CONFIDENTIAL, pages numbered	Must be marked RESTRICTED	Marking by data owner discretion	No requirements
Copying of documents	Not allowed	Data owner authorization required	Copies for legitimate purposes allowed	Copies for legitimate purposes allowed	No restrictions
Mailing of documents	Doubled enveloped, with no outside security marking; Signed receipt	Doubled enveloped, with no outside security marking; Signed receipt	Sealed envelope with no security marking	Sealed envelope with no security marking	No requirements
Transmission by Fax	Encryption required	Encryption required	Encryption optional; data owner to specify	Unencrypted	Unencrypted
Review of classification	Review date to be specified by data owner; Consider downgrading	Review date to be specified by data owner; Consider downgrading	Review date to be specified by data owner; Consider downgrading	Review date to be specified by data owner	No requirements
Destruction of documents	Controlled physical destruction required	Controlled physical destruction required	Controlled physical destruction required	Controlled physical destruction required	No requirements

COMPANY COPYRIGHTED MATERIAL

At regular intervals, some of your fellow employees will be creating new work in the form of application programs, transactions, systems, and so forth. To protect your organization from the loss of created material you must ensure that all employees understand the copyright law and how it should be applied to their work.

Material Protected By Copyright

Unlike other forms of intellectual property protection, the basis for copyright occurs at the creation of an original work. Although copyrights are granted by the U.S. Copyright Office, every original work has an inherent right to a copyright and is protected by that right even if the work is not published or registered with the U.S. Copyright Office. Copyright registration is, however, required before legal action can be filed for infringement of copyright.

All original works of authorship created by employees for a company are the property of the company and are protected by the copyright law. The copyright also applies to suppliers doing work for your organization while under a purchase order or other contractual agreement. Unless there is an agreement to the contrary, any work created by a contractor under a purchase order to your organization is owned by your organization, not the contractor.

The types of work that qualify for copyright protection include:

1. All types of written works.
2. Computer databases and software programs (including source code, object code, and micro code).
3. Output (including CRT screens and printouts).
4. Photographs, charts, blueprints, technical drawings, and flowcharts.
5. Sound recordings.

A copyright does not protect:

1. Ideas, inventions, processes, and three-dimensional designs (covered by patent law).
2. Brands, products, or slogans (covered by trademark law).

After a work has been created, due care must be taken to preserve copyright ownership. Your organization's rights to its copyrighted material may be reduced or even lost if the material is published or publicly disclosed without complying with the copyright notice requirements.

Copyright Notice

The copyright notice informs others that the material is a protected, copyrighted item of your organization and that it shall not be reproduced without written permission. To be legally valid, the notice must contain certain elements, and the specific elements depend on whether the material is published or unpublished. The notice may also include additional information about the material (such as the Standard Book Number or SBN) and whom to contact for permission to reproduce. It should be noted that the copyright notice does *not* relieve your organization from the requirement to keep the material properly protected from unauthorized access.

The following are the required elements for proper copyright notice and some examples of the notices:

1. For published works (material that is intended or likely to be distributed outside your organization), the notice must contain the following three elements: the symbol "©" or the word "Copyright," the year that the material was first published, and the copyright owner's name.

Examples of copyright statements for published material are:

© 1991 Commonwealth Films, Inc. All rights reserved.

© 1991 Commonwealth Films, Inc. All rights reserved. This material shall not be reprinted in whole or in part without the express written permission of Commonwealth Films, Inc., Boston, MA.

2. For unpublished works (material that is not intended to be disclosed or used outside your organization, for example, custom computer programs) the notice must contain the following two elements: the symbol "©" or the word "Copyright" and the copyright owner's name. The year is not included since the material is not considered published.

The notice may also contain the name, address, and phone number of the division, group, or department that created the material as well as the current status or other appropriate information about the material.

Example of copyright statements for unpublished material are:

© Miller Freeman Publications, a member of the United Newspapers Group

© Miller Freeman Publications. This is unpublished material created in 1991. It shall not be copied, distributed, or otherwise used outside of Miller Freeman without the express written permission of Miller Freeman, San Francisco, CA.

The following notices may be used for classified material:

© Commonwealth Films, Inc. This unpublished company confidential material of Commonwealth Films was created in 1991 and may be copied and used only by Commonwealth employees within the scope of their employment. It must not be copied or distributed outside Commonwealth, in whole or in part, without the express written permission of _____.

© Micro Security Devices (MSD). This is unpublished, restricted, MSD material created in 1991 and is being provided to MSD suppliers solely for use in conjunction with the work being performed under a MSD purchase order. It must not be disseminated or disclosed further or additional copies made (except for necessary backup purposes) without the express written permission of _____. Upon completion of the work performed under the MSD purchase order, all copies must be returned to MSD.

Location of the Notice

The copyright notice should appear in a location that gives reasonable notice of the owner's copyright claim. As a practical matter, you should place the notice at the beginning of the material. For example, with written material, the notice may be placed on the title page, the page immediately following the title page, or the inside front cover. For computer software programs, the notice may appear on any of the following: on JCL listings and printouts; at the user's terminal during sign-on; continuously on the terminal display; as a comment in the source code; or on a label securely affixed to the tape, cartridge, disk, or storage device holder.

FIGURE 6 **AWARENESS POSTERS**

INFORMATION INTEGRITY

Information integrity is another important element in an information security program. Information integrity can be defined as the assurance that the information used in making business decisions is free from inaccuracies. Integrity of information is often taken for granted, but Figure 6 points out that integrity is far from guaranteed.

For most employees, the concept of information classification is generally understood and accepted as a positive business management tool. This issue of information integrity, however, is not intuitively clear. Users generally accept computer printouts as correct. The days are long past when accountants would sit and add the figures on the computer printout to verify the totals. Engineers no longer take out their slide rulers or calculators to see if the results from the computer are acceptable. Users have abandoned the healthy skepticism that kept data processing departments on their toes. And the security steps taken over the past two decades to protect the computer facilities and the data in them have added to the acceptance level. In the meantime, the introduction of microcomputers has resulted in erosion of protection in ways not envisioned previously. As a result, you must ensure that your information security program allows changes to data only in a controlled and structured environment.

An organization's policy on data integrity should take into account the practical day-to-day operation of the computer systems throughout the organization, the classification value of the data, and the probability of risk to data integrity. Integrity controls ensure that the established procedures are adhered to by the users and are supported by the system.

To implement the best control procedures for the integrity of information and computer systems, there are three basic principles that must be considered.

Basic Integrity Principles

Each employee with access to the computer system or the organization's information should be assigned a specific set of functions, privileges, restrictions, and capabilities. The overriding principle of *least privilege* should govern this process. Under least privilege, the employees or users are assigned only the minimal level of access required to perform their specific job function.

Some practical examples of least privilege are:

- An application programmer must not have the ability to update production source code directly. As a matter of fact, an application programmer should not have access to any production source code without an active, current service request from users and an approval from the programmer's manager.
- Classified data should not be placed in publicly available files, data sets, or databases. When using an access control package, the universal access code for classified data should be set to "NONE."
- Employees should not have access to the criteria used by your organization to trigger an audit investigation.

A perfect example of an access violation is discussed by Buck Bloombecker in his book, *Spectacular Computer Crime*. Two bank employees (one current, one former) were able to embezzle large sums of money by understanding the parameters that would cause an audit investigation. By ensuring that the amounts they embezzled did not exceed acceptable levels and by working within certain time parameters, they simply issued cashier's checks to cover their needs.

As Buck points out:

> If you cannot keep employees from gaining access to valuable information, . . . you can at least keep the information they get from remaining valuable. Good security procedures are usually designed from the premise that 'the enemy' will eventually learn what the procedures are. Thus changing the

procedures or varying the triggers that cause the security procedures to go into effect are two ways business can reduce its vulnerability to employee computer crime.[1]

All employees must be granted sufficient access to perform their job assignments. For unclassified data or even some types of restricted data, some group or departmental access may be reasonable. At higher classification levels, however, access must be granted only to individual users. The key to control is to maintain individual accountability.

As with any other control mechanism, the principle of least privilege must be tempered by the practical concerns of the business environment. Alternatives always exist that will allow the organization to establish maximum controls with minimum impact on daily work.

The second basic principle of information integrity is that of *separation of duties*. Using this principle, no single employee should have complete control of a transaction from inception to completion. Some examples of the separation of duties are:

- Employees having the ability to create a purchase order, edit the receiving system to indicate that the materials have arrived, and then issue the disbursement check for the purchase order. At each level, adequate controls must be in place to ensure that the organization is not operating at risk.

- Administering the organization's access control authority lists and then auditing them. As stated previously, the owner of the information is the only individual authorized to approve additional use of the resource. While the owner may not be the employee who will write the rule or issue the permit to access, he or she is charged with the responsibility to monitor access control lists. The access control administrator should not be in charge of the auditing function.

Controls can be worded in many ways. A dogmatic control might be stated as follows: "Individuals who authorize purchase orders are not permitted to authorize the payment checks." Although this is the desired level of control for most organizations, it may leave the local unit with little latitude. A more flexible solution might be: "Employees can authorize both purchase orders and payments, but no individual should authorize both the purchase order and the payment check for the same item." If the control is still too restrictive for some locations within the organization, an additional option might state: "In those facilities without sufficient personnel to enact the control mechanism, the employee responsible for auditing the accounting activities must not be the employee responsible for the day-to-day activities."

The third basic principle of information integrity is the *rotation of assignments*. There are always some assignments that can cause the organization to be at risk unless proper controls are in place. To ensure that published procedures are being followed, employees should be assigned to different tasks at intervals throughout the year.

Some organizations maintain that rotation reduces job efficiency. However, it has been proven that an employee's interest declines over time when the employee stays in the same job. Also, employees sometimes take shortcuts when they have been at a job too long. Some managers resist the idea of rotation because they feel that their department runs more smoothly when they aren't being asked a lot of questions about how to do a particular job. On the plus side, every time an employee is given a new assignment, the organization can review how the job was being done in the past and see where changes could be made.

In general, separation of duties is designed to ensure that fraudulent practices require the collusion of individuals. The rotation of assignments is a complementary principle that removes one of the colluding parties from the task, thus exposing the other(s) to detection. Financial institutions often require employees to take vacations or even attend off-site classes for two consecutive weeks as a way of rotating assignments. For all businesses dealing with sensitive data, attention must be paid to employees who refuse to take allotted time off.

INFORMATION AND SYSTEM AVAILABILITY

Another essential element in an information security program is the availability of information and systems that are integral to doing business. The information owner (discussed at length later in this chapter) is responsible for identifying critical assets or those assets that would severely affect the operation of the organization if they were unavailable. In evaluating whether assets are critical, consideration should be given to the processing hardware, system software, applications program, and essential human resources. For systems identified as critical, the system owner should stipulate the duration of acceptable system outage, and appropriate control procedures should be coordinated with the data processing department's disaster recovery plan.

Availability is the initial phase of your organization's disaster recovery plan (and will be discussed at greater length in Chapter 4). Briefly, the availability requirements for a specific application or system must be identified by the owner. This information will be used in the critical application process. This process is used by the disaster recovery plan coordinator (refer to

Figure 7) and management to assign recovery priorities. That is, the more critical the system or application to the business, the sooner the system must be recovered.

The goal for your computer security policies and procedures is to ensure that the data processing requirements are included as part of your organization's overall business resumption plan (BRP). The data processing requirements are just one of many that make up the BRP for your organization. As you can see in Figure 8, the data processing requirements must be coordinated with the total BRP.

FIGURE 7 **BUSINESS RESUMPTION ORGANIZATION CHART**

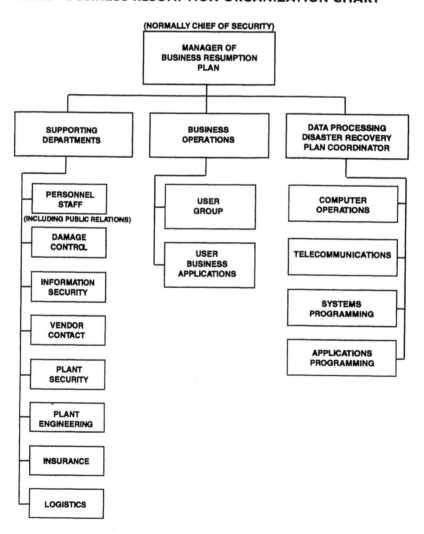

FIGURE 8 **DISASTER RECOVERY PIE CHART**

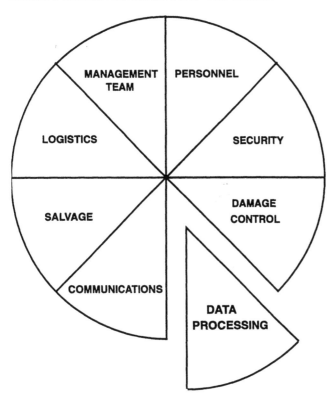

CARE AND CONTROL OF INFORMATION

The final essential element of an information security program is to establish clearly defined lines of responsibility. As discussed in Chapter 1, senior management is ultimately responsible for protecting the organization's assets. To protect assets, the role of the owner, user, custodian, and auditor must be defined. The following definitions are currently in standard use in the information security industry.

Owner. Ownership is the exclusive right to use. Information owners are the senior-level managers who have been formally assigned to exercise the organization's proprietary rights and fiduciary responsibilities for the information. The information owner is responsible for the following:

1. Judging the value of the information and assigning a proper classification to it.

2. Authorizing access to the information.

3. Specifying control requirements and communicating these requirements to the information custodian or steward.

4. Performing periodic reviews of controls to ensure that information is only available to persons with a specific need, data is free from significant risk of undetected change, and a disaster recovery plan is available if necessary.

5. Reviewing and reconciling data access security violations and log reports.

Additionally, the designated owner is responsible for assessing the degree of risk involved with ensuring information integrity and availability and determining if adequate controls are in place to mitigate such risks.

The information owner may delegate these responsibilities to another individual (for example, department manager to department employee); however, these responsibilities are not to be delegated to the steward or custodian.

Users. Users of information have access to it under the supervision of their managers. Users are responsible for protecting information in accordance with the directions of the information owner.

Being granted access to secret, confidential, or restricted information does not imply or confer authority to grant other users access to that information. Only the information owner has the authority to grant access. Therefore, if granted access to secret, confidential, or restricted information, users must seek the authority of the information owner before viewing such information.

Users authorized to access the organization's information are responsible for:

1. Using the information only for the purpose intended.
2. Complying with the established controls.
3. Not disclosing information to anyone without written authorization from the information owner.
4. Providing adequate control over information in their possession.

Custodian/Steward. Custodians or stewards of information hold it for the owners in accordance with their directions. Custodians or stewards grant access to the information users who have been authorized by the information owner.

The custodians or stewards are responsible for the following:

1. Providing physical safeguards for mainframe and network environments.
2. Providing procedural safeguards.
3. Administering access to information when approved by the information owner.
4. Implementing cost-effective controls.
5. Providing recovery capabilities.

Auditor. The internal audit staff is responsible for performing periodic security reviews of the organization's computer facilities, personnel, data, and software. These reviews are used by senior management to prepare the certification statements published in the annual reports. The audit reports show senior management that the organization is in compliance with established policies and procedures, that any areas of noncompliance have been identified, and that corrective actions have been implemented. It must be noted that auditors do *not* write the policy; their role is to check compliance. However, where there are no policies and procedures, auditors are required to perform their reviews based on currently acceptable industry standards.

Some additional or alternative roles are:

Guardian. An individual or organization responsible for the accuracy and integrity of the data, programs, and systems.

Supplier of Services. Data processing suppliers outside the company are required to provide adequate physical and procedural safeguards to ensure that access to classified information stored in their computers (including all magnetic media) must be controlled as prescribed by the owners of the data.

AUTHORIZATION FOR ACCESS

The protection of the organization's assets, whether physical or intellectual, is a basic management responsibility. Managers must identify and protect all assets within their assigned area of management control. Managers are responsible for ensuring that all employees understand their obligation to protect the organization's assets. Managers are also responsible for implementing security practices consistent with those published by the organization. To establish clear lines of authority, some key security responsibilities must be assigned to senior management, data processing management, user management, and audit staff, as shown in Figure 9.

FIGURE 9 **AREAS OF RESPONSIBILITY**

Senior Management	Data Processing Management	User Management	Audit Staff
Establish security goals & policies	Issue computer security procedures, practices, etc.	Ensure employees are aware of security controls	Review compliance with policies and procedures
Establish computer security function	Monitor security function	Educate employees on security compliance	Monitor employee information security awareness
Establish a security steering committee	Chair security steering committee	Classify data and systems based on procedures	Review data classification process
Integrate business and systems development plans	Explore use of automated tools to improve business performance	Establish an effective records management program	Monitor records storage and destruction requirements
Review and approve systems developement and maintenance projects	Monitor systems and application development and maintenance projects	Notify data processing on backup requirements	Monitor data backup program
Establish organization-wide business resumption plan	Ensure user backup and storage requirements are met	Develop business resumption plan to support data processing DRP	Review DRP
	Develop data processing disaster recovery plan (DRP)		Notify senior management of current security compliance conditions

CHAPTER REVIEW

1. Classification of information establishes the security control levels necessary for appropriate protection of the resources.
2. The four criteria used to legally determine if the information is confidential are:
   - The information provides a competitive value.
   - The information was the result of some expense.
   - The information is not commonly known.
   - The information is accessed by those employees with a clear business "need-to-know."
3. Employees with access to the highest classification level might be required to sign an employee agreement.

4. A program of declassification and destruction must be implemented.

5. Company material must contain the appropriate copyright notices.

6. Information integrity is an essential element of an information security program. Integrity ensures the quality of the systems, programs, and data.

7. The three basic principles of information integrity are: least privilege, separation of duties, and rotation of assignments.

8. Availability is also an essential element of an information security program.

9. Establishing the responsibility for care and control of data is an essential element of an information security program. The owner, user, and custodian of the information must be defined.

10. The protection of the organization's information is a basic responsibility of management.

Exercise 1

Using the material discussed in this chapter, create data classification levels for your organization. Remember to use your organization's naming conventions whenever possible. Use the examples given in this chapter as a guideline for your classification program.

Exercise 2

Using the blank matrices provided in Figures 10 and 11, develop your own control matrix for data classification. Be sure to include classification for data stored in a computer and for printed data.

Exercise 3

Employee Conduct Policies and Procedures
- Define sensitive job functions and procedures.
- Require separation of duties to prevent fraud from being perpetrated by an individual employee.
- Establish disciplinary measures to be taken upon detection of fraud or theft or destruction, modification, or disclosure of data.
- Clearly define limitations of employees' personal use of company resources—include computers and other equipment as well as supplies, manuals, and documentation.
- Describe who can access what sensitive data. Include all the clearances to be instituted.

FIGURE 10 **BLANK MATRIX: COMPUTER DATA**

FIGURE 11 **BLANK MATRIX: PRINTED DATA**

• Define employee responsibility in reporting any known or suspected misuse of organizational resources.

• Specify required termination procedures.

CHAPTER NOTES

1. Bloombecker, Buck, *Spectacular Computer Crimes* (Homewood, IL: Dow Jones-Irwin, 1990), p. 92.

CHAPTER 4

Volume Contents

As described, security is primarily a people problem, not a technical problem. Therefore, the security solutions you develop must address the way employees work. If security is a problem at your facility or is a new concept, security steps will have to be implemented within the framework of the existing organizational structure.

As a people problem, security is actually a management issue. Management sets the atmosphere for the development and acceptance ·of the controls required to ensure a constructive security environment. Chapter 6 discusses how to get management involved in the security program. For now, just remember that management must become an active player in the security process or the security program will stall and may eventually fail.

LEVELS OF RESPONSIBILITY

Each level within the organization provides specific support for the security program. Management is responsible for creating and implementing all administrative policies and procedures for security. These procedures apply to all levels of security, of which the information and computer security issues are a subset.

Within data processing, the areas of responsibility are:

Technical staff—Provides all systems development, maintenance, and applications development standards. This group includes the systems programming, systems analysis, and applications programming staff.

Operations staff—Provides most physical security controls for the computer facility, such as disaster recovery planning, data backup, off-site storage, hardware installation, production job scheduling, tape library management, and Direct Access Storage Device (DASD) maintenance, among other tasks.

Security staff—Administers the access control package, monitors audit trail reports, administers userid and common user data sets (SYS1.SHRPROC, SYS1.CMDPROC), and develops computer and information security policies and procedures.

The users are responsible for security within their own work area, including requirements for microcomputers, either stand-alone or networked. Microcomputer requirements would include such controls as macro code review, data set backup, disaster recovery planning, access control (both to the unit and to the data on the hard drive), documentation development, and so forth.

TYPES OF CONTROL

Controls should never be put into place just for the sake of having controls. There should always be business objectives that will be met by the controls. Because every organization has its own character, different types of control will work differently for each organization. Nevertheless, there are some generally accepted types of control in the industry; they include management, preventive, containment, and recovery controls. Each type has its use and can be effective, but no one type of control is universally effective, and the best results are often achieved by using a combination of these types.

Management controls consist of actions taken by management, in advance of a security breach, to encourage compliance by employees. Management controls are a series of preventive measures to ensure that the business will function with consistent controls.

The initial phase of management controls involves establishing a security steering committee. This committee is normally made up of the various data processing departments and user groups, and should also include representatives from the audit and security staffs. The committee's function is to approve the goals and objectives of the security program, including the security mission statement.

Management must give full support to the security function by approving the mission statement. This empowerment is a vital element in the organization's acceptance of the security controls. Additional security controls are:

- Assigning asset responsibility and accountability to specific employee classifications.
- Approving and providing the means for distributing security policies and procedures.
- Supporting the creation of an employee security awareness program and endorsing funding for the program.
- Providing guidelines to be used by employees for classifying and protecting the organization's assets.
- Creating a software code of ethics for copyrighted products, both those purchased and those developed in-house.
- Establishing guidelines for the control and implementation of systems development and maintenance.

• Establishing guidelines for software coding and ensuring that all affected employees are properly trained in the guidelines.

• Establishing criteria for user testing and acceptance of developed software.

• Establishing a security review process to certify department, staff, section, unit, and division compliance with security requirements.

• Initiating and approving standards or policy changes and additions.

The following sections describe types of controls that help to fulfill the four main objectives of a security program. As you remember, no security program can be 100 percent effective. With that level of security, most organizations would cease to function. If security cannot be totally effective, it might be argued that the security program should at least prevent security problems—but it's obvious that there will always be some problems. Given that there will always be some problems, the goal of the security program is then to detect a problem, contain the situation, and provide the capabilities to recover from a problem.

Preventive controls augment the controls put into place by management. The first step of preventive control is to make sure the physical plant is secure. In order to secure the work site, access requirements must be established. A program of identifying employees, contractors, visitors, and others must be set up, so that only authorized personnel have access to the buildings. This program would also include the inspection of packages brought into and taken out of the facility and securing access to telecommunications storage areas and control rooms.

Once controls have been instituted to secure the physical environment, system access must be addressed. System access can be restricted by implementing access control lists with userid and password controls. For employees and authorized system users requiring remote system access, an additional level of security must be implemented. This control normally takes the form of either a system dial-back or a token password requirement. After logging on to the system, users should only be granted access to the information required to complete their assignments.

Preventive controls also include encryption of data being stored on the system or transmitted to a remote location. Document distribution requirements will have to be established, and data will have to be reviewed on a regular basis to ensure that controls are still required. Finally, the data, information, data sets, documents, and other materials will require proper destruction when they are no longer required or when they have reached the legal limit for retention.

Establishing preventive controls for the programming areas will be your greatest challenge. Since the programming areas have the greatest potential for loss or exposure, controls must be established for creating new high-level qualifiers, which will go hand-and-hand with a standard naming convention system. Along with management controls on software coding, preventive controls include structured programming requirements. Additionally, there must be controls to make sure that programmers do not have access to source code without proper authorization and that the programmers do not have the ability to directly update or recompile programs.

For systems programmers, access to ultra powerful userids (such as Special under RACF, Non-Cancl under CA-ACF2, Hot-Keys under CA-TOP SECRET, Supervisor under VAX, and Superuser under UNIX), must be tightly controlled. While an emergency userid will be necessary, its use must be monitored and recorded. In addition, powerful utilities, such as Superzap, should require authorization and must be audited. All system modifications must be reviewed prior to implementation, and all requests for change must be approved by any affected parties. The system programmer, as part of the request, should have a back-out procedure documented.

Finally, there must be a well-established division of work responsibilities. Separation of responsibility (as discussed in Chapter 3) must be implemented throughout the data processing departments.

With management and preventive controls in place, the job of the security officer and auditor now begins. Eventually, automated *detective control* programs can be used to provide feedback on whether the preventive controls are succeeding. By using the audit trails provided and reviewing system logs and error execution reports, the security officer and auditor can judge the level of compliance to the established controls.

The remaining detective controls fall into the category of audit requirements. Auditors will often ask to review the logs for physical controls. To satisfy auditing requirements, procedures will have to be developed to inventory all hardware and software and to keep the list reconciled and up-to-date. The inventory should include the tape library and card access system.

- The security officer will also be audited to ensure that system activity logs are reviewed and reconciled on a regular basis and that after-hours security checks are conducted to make certain that information is properly stored, that PCs are locked up, and that terminals do not have system log-on information displayed.

In the programming area, procedures should ensure peer review of new source code. For the normal program, peer review usually covers 10 percent of the source code. For critical systems or confidential data, the review should cover all of the source code. Annual inventories should be conducted to compare source code to object code, and documentation should be reviewed to determine that it is current.

Despite an ambitious program of preventive controls, you will still have security breaches. *Containment controls* provide procedures, both manual and automated, that will identify errors, irregularities, or omissions and limit the damage to the system and the organization.

As with all controls, the goal is to have the controls in place before a problem occurs. Containment controls include the use of structured programming for all application development work. Often organizations bring in or hire new programmers to complete a job assignment, typically when the project is behind schedule. The programmers are then rushed into production with little or no training on how the organization expects code to be written.

To assist in this process, procedures for using ABEND software monitors for program debugging must be standardized and in place. While time is often a deterrent to structured programming, requirements for using "vernacular-like comments" in the source code and JCL must be enforced. A formal walkthrough should be conducted for all application and system development projects. A programmer peer group or development team should also review source code, listings, output, and other programming aids. Finally, the customer must be actively involved in reviewing, testing, and approving the product. Testing should include a review of audit trails and ensure that the program checks for the reasonableness of the input.

While containment controls serve to limit a security breach, *recovery controls* are used to resume operation after a security breach. Recovery controls often consist of a disaster recovery program and include determining that requirements for data backup and off-site storage have been defined by the owner department, transmitted to data processing, and monitored by all parties for compliance. All departments, including data processing, must develop and test disaster recovery plans to make certain that duplicate copies of documentation are stored off-site, and that forms, supplies, and other office materials are either sorted off-site or arrangements have been made with suppliers to make them readily available when necessary.

Emergency response procedures must also be developed and employees must be made aware of them. These procedures must be reviewed and tested with the employees on at least an annual basis. Never assume that employees will leave a burning structure or, if they do leave, that they have turned on the emergency alarm. It may come as a shock to the MIS director when she returns from lunch one day to find all the operations employees in the parking lot, none of them having notified security of the emergency in progress.

COST-BENEFIT ANALYSIS

When preparing a control policy, your organization must conduct a cost-benefit analysis. Ideally, the controls will provide maximum security for an acceptable investment. However, cost should not be the only determining factor.

Another factor that should be considered is that security controls help maintain the integrity of the data, information, applications, and systems for your organization. As discussed previously, the key element in security is the integrity of the information used to make business decisions. Therefore, while cost is often the first factor that is considered when implementing a security program, it is by no means the most important factor.

Another benefit of security controls is improved customer service. All data processing facilities are in business to serve the needs of the users, but it's easy to lose sight of that goal. If your customer base were another organization or company, your desire to improve customer service and system response time would be greater than it is when your customers are only your fellow employees. However, the needs of your customers, regardless of who they may be, should be a major factor in any cost-benefit analysis.

Timing should also be considered in any discussion of cost-benefit analysis. Is the timing right for increased controls and security? Users may be hesitant to accept some new data processing controls if there has just been a traumatic system upgrade or some other disruption in service. To many users, the phrase "transparent to the user community" translates to "they never saw it coming." Because timing is critical, make sure you prepare the users to accept any new control mechanism and allow them time to get accustomed to a new procedure before going full speed ahead with an array of controls.

A cost-benefit analysis should also include the prospect of increased security. The security concept is becoming less and less of a hard sell. Most organizations are aware of the increased threats to the organization and are willing to invest in developing security procedures. Most employees will also accept the concept of security, if they are first made aware of the need.

IMPLEMENTATION OF CONTROLS

One obstacle to your progress toward a security program is that most organizations are already established and have existing systems in place. No matter what you develop and no matter how logical the procedures, you will have to retrofit the new requirements to the existing systems. Adding security to a mature company generally costs, on the average, four times as much as it does to design security during the development phase of the organization. By recognizing the obstacles to retrofitting before you start, you will be able to provide effective solutions.

As you initiate controls, try to create a baseline, so that at some point in the policy and procedure approval process, all projects under development will have to adopt the new standards. To make a smooth transition to the new controls, you should establish a liaison with the development personnel. Then development personnel will be aware of the new requirements and can begin to incorporate the controls into new projects. Do not wait until all the procedures have been approved and published before you meet with the systems analysts; they must be involved in the process as soon as possible.

Implementing controls incrementally will save your organization time, money, and frustration. By working with development and other personnel, your procedures and policies will be accepted and implemented more readily. Well-designed and integrated policies will improve the audit process by ensuring adequate controls. Compliance will result in improved security and protection for the organization's assets, while lowering the risk factor. Finally, the procedures will be easier to maintain and update.

PROCEDURE CONTENTS

The policy statement developed in Chapter 1 must now be translated into the procedures, practices, standards, and guidelines that will support the security objectives. Good procedures, as you recall, must address the organization's business objectives and be cost effective.

When you research procedures developed by other organizations or attend conferences or seminars, you will be inundated with ideas. All too often, attendees go back to their own organization and try to achieve the level of security in effect at companies that have developed security procedures over many years. Remember, movement to controls is an evolutionary, not revolutionary, process. Your procedures must reflect the security expectations at your organization and must be appropriate to your organization's style. Good procedures offer practical solutions to the problems of asset production.

To enlist the support of the various departments within your organization, try to make your procedures as complete as possible. All procedures will have to be reviewed and modified over time, of course, but you should start out by making them as comprehensive as you can.

Over the years, a pattern of volume contents has come to be accepted as minimum industry guidelines. Unlike other industries, the security industry has no uniform set of standards. A number of industry organizations are attempting to establish minimum standards, but the results are still incom-

FIGURE 12 **VOLUME CONTENTS**

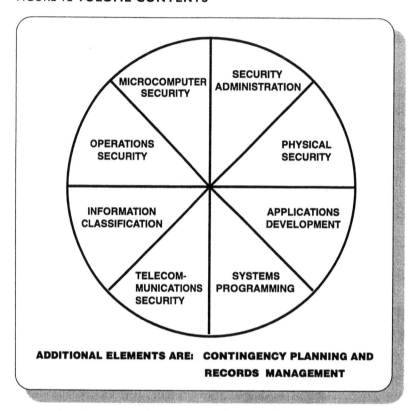

plete. To ensure that your security procedures meet a minimal level of security, you will want to develop at least eight and possibly ten chapters of procedural material in your manual. Figure 12 identifies the basic security sections to be addressed.

As you may have noticed, contingency planning and records management are not part of the pie chart. You will want to include these two topics in your manual, and it is certainly acceptable to devote a separate chapter to each topic, if you wish. However, each section of the pie chart must incorporate minimum requirements for records management and contingency planning.

The remainder of this chapter will discuss the specific areas that should be included in your organization's security manual. Each section will be addressed individually and the objectives described for that section. Each section will include a sample policy statement, the minimum procedural requirements, and an example of a procedure.

PHYSICAL SECURITY SECTION

Security programs are often instituted to satisfy a specific need—an actual loss or exposure or a deficiency identified in an audit. When developing a computer and information security program, physical security is normally addressed first, partly because physical security is the easiest to understand and because it provides the greatest opportunity for showing tangible results.

Physical security for a computer facility usually consists of the physical plant, access control, and emergency procedures. For the physical structure, you must determine if the facility meets established standards for safe computer operations. These standards include federal, state, local, and organizational requirements. Along with the physical structure, environmental requirements should be addressed, such as heating, cooling, and humidity and the condition and reliability of the electrical system.

Access control to the computer facility is another element of physical security. Many organizations have created special areas where the computers are partitioned off from the people who run them. The rooms meet the desired environmental standards for computer operation and normally have motion sensors that turn out the lights when no humans are in the area and notify security when movement is detected. The human room (starting from the outside to the inside) consists of the following areas: input/output (I/O), tape management, and council area. Each area has its own access control procedures and authorization levels.

Emergency procedures are another element of physical security. With emergency procedures, harm to employees, equipment, facilities, and so forth is kept to a minimal risk.

Physical security overview statement

Physical security procedures must be implemented and enforced to prevent unauthorized access to buildings, computer facilities, and equipment. Only authorized personnel with an established business need should be permitted access to the computer facility, and all visitor access must be approved by the appropriate computer center management. Furthermore, visitors should be escorted throughout the facility during their stay.

Housekeeping controls must be in place to ensure that environmental conditions (heat, humidity, dust, and so forth) are adequate to prevent the possible loss of processing services.

Physical security minimum requirements

To complete your minimum requirements, you should include individual procedures on the following topics:

- Employee identification process
  - Authorization for access
  - Visitor access requirements
  - Vendor and maintenance employee access
- Physical plant controls
  - Computer center
  - Control unit rooms
  - Microcomputers
  - Remote job entry locations
- Fire control
  - Detectors (installed and tested)
  - Extinguishers (types required and tested)
  - Employee fire brigade training
  - Emergency notification and education
  - Evacuation procedures
- Inventory of equipment
  - Hardware
  - Software
  - Tape, cartridge, and other magnetic media
  - Terminals, control units, cables, bus, and tags
- Electrical requirements
  - Annual review of supply requirements
  - Filtering capacity
  - Uninterrupted power supply (UPS)
  - Diesel backup requirements
- Housekeeping requirements
  - Normal maintenance
  - Filter changes
  - Dusting requirements
  - Smoking within the computer room
  - Food and drink controls
- Vital physical requirements (contingency plan)
  - Computer facility blueprints
  - Electrical diagrams
  - Air conditioning requirements
  - Equipment inventories
- Guidelines for records disposal
  - System logs
  - System Management Facility (SMF) records

Information Management System (IMS) log tapes

Access control package error reports

Customer scheduling requests

Removal of backup tapes from storage vaults

PHYSICAL SECURITY SAMPLE PROCEDURE

Implement an effective physical security program.

Physical security and access control are the first line of defense and the most effective means of preventing unauthorized personnel from entering the computer facility.

The following are minimum requirements:

1. All employees are required to wear their identification badges when on the premises.
2. All visitors are required to: sign-in on the visitor log; be approved by the authorized computer center management; be escorted throughout the facility.
3. Access to the computer facility will be controlled via card access system.
4. Emergency exit doors are to be equipped with alarms.
5. Access to the intensive control zone will be controlled by card access and turnstile.
6. All packages, briefcases, lunch boxes, and other items are subject to inspection at any time.

SECURITY ADMINISTRATION SECTION

Most information security officers appreciate the fact that their job is subject to the same controls as all other employees. Some security officers, however, will resist controls so it is important to establish limits and areas of responsibility for the computer security area, just as you have with other departments in the organization. Additionally, the security administration procedures actually set the tone for the way other users interact with data processing.

Security administration overview statement

As data processing becomes integrated into the general office environment, adequate controls must be established to reduce the potential of unauthorized access or modification or destruction of the organization's data and systems. To achieve control, access privileges should be established and administered in accordance with the owner's authorization requests. Data file creation, maintenance, and monitoring should be performed in compliance with established procedures.

Security administration minimum requirements

The goal of security administration is to limit the chance of exposure and risk to the organization's assets. The requirements include evaluating potential security breaches, providing legitimate access to the computer system for all authorized employees, and establishing programs to remove invalid userids, data sets, and programs from the system. Security administration policy should also ensure that the computer facility cannot be easily identified by exterior building markings, internal identification, or listing in the company telephone directory.

Additionally, the terminal log-on screen should not identify the company or provide any unnecessary information. Legal rulings have stipulated that a greeting such as "Welcome" is an invitation to access, so by all means eliminate it from any terminal screen. Also, if you include a threat of prosecution on your sign-on screen, your organization must be willing to prosecute every infraction, otherwise you will be charged with selective enforcement and truly don't have a security program. An effective terminal screen currently being used by a number of U.S. companies is shown in Figure 13.

Basic security administration procedures should address the following:

• Access control package administration

    Userid request

    Access list development

    Universal access requirements

• Employee userid and password program administration

    Minimum password length

    Number of password histories

    Authorizations required for access

    Automated password changing requirements

FIGURE 13 **SIGN-ON SCREEN MESSAGE**

WARNING

THIS SYSTEM and DATA HEREIN ARE AVAILABLE ONLY FOR AUTHORIZED COMPANY WORK. USE FOR ANY OTHER PURPOSE MAY RESULT IN ADMINISTRATIVE OR CRIMINAL ACTIONS AGAINST THE USER.

Employee generated password versus assigned passwords

Notification of userid change in status

• System access terminal requirements

Establish automatic log-off for inactivity

Establish threshold for unauthorized access attempts

Implement user log-on controls to prevent or detect improper or invalid attempts

• Log monitoring

Review and reconcile system activity logs for exception reports

Review access control package logs

Notify data owners of unauthorized access attempts

• System maintenance requirements

Perform DASD maintenance on a regular basis

Establish system backup requirements

Establish automated sequence for obsolete userids

Establish regular reviews to ensure continued necessity for data sets

• Contingency planning requirements

Backups are stored off-site as required

Access to off-site storage is restricted

Disaster recovery plan section is current and off-site

• Records management requirements

System and userid authorization requests

Reconciled system logs

User incident reports

SECURITY ADMINISTRATION SAMPLE PROCEDURE

Authorization for Employee System Access

Employees requiring access to the computer systems must be approved by their management prior to receiving a user identification (userid). A completed, authorized request form or an electronically-generated request must include the following information:

1. Employee name
2. Employee social security number (not required in Canada)
3. Employee number (not required in U.S.)
4. Department name and code
5. Address
6. Phone number
7. Department billing code
8. System to be accessed

Automated access control systems that incorporate unique identification and authorization (such as a userid/password combination) should be used to access the computer systems. The following are system access control minimum requirements:

1. If an automated access control system is not feasible, appropriate compensating controls must be used.
2. Each user of the system should have an individual userid.
3. Procedures should ensure that the userid is removed from the system when the employee is terminated, transferred, or no longer requires access.
4. Users should be required to log-off or secure the terminal when exiting.
5. An automatic timeout or reauthorization should be required after a specified period of no terminal activity.
6. The user should provide authorization (such as a password) that is known only to that user.
7. Passwords should be selected by the user.
8. Passwords should never be given to another individual.
9. Passwords entitling users to access computer systems should be changed at an interval not to exceed ninety days.
10. System software should disable the userid after a given number of consecutively invalid password attempts.

APPLICATIONS DEVELOPMENT SECTION

The development of application programs and computer systems is vital to most organizations. To meet the needs of the organization and develop applications in a timely and controlled manner, procedures must be established and communicated to programming staff. As the demands for information integrity grow, quality assurance takes on increasing importance. Procedures on programming parameters will ensure that application development work is completed within an acceptable time frame and that maintenance on a program or system will not require a specific programmer.

Applications development overview statement

An organization's business objectives can only be met by effective, computer-based information systems. It is, therefore, essential that the development of new information systems and major enhancements to existing systems be carefully managed. Developing new systems and enhancing existing systems are complex tasks involving many people over extended periods of time, and they must be executed in concert with the organization's written procedures.

Applications development minimum requirements

Most organizations have established or should be able to establish financial control over application development. The appropriations manual should identify the authorizations required when new system development is being considered. These authorization levels are normally associated with increasing levels of expenditure. Before any money, time, or effort is spent, there must be some form of authorization to begin a business case study (BCS). The BCS will examine existing applications to see if they meet the new requirements, what business need will be served by the new application, and, finally, the estimated cost of the project. If these approval requirements do not currently exist within your organization, they will have to be developed and included as part of the applications development section.

The applications development section should also cover the following items:

- Request for service
    Authorization before beginning application development
    Establishing an owner for all required approvals
    Customer involvement in the development phase
    Customer participation in the testing phase
    Final customer sign-off upon completion
- Programming standard methods
    Structured programming methods
    Uniform coding requirements
    Data dictionary
    Naming convention requirements
    Comment cards in source code and JCL
    Automated code optimizers, tracing tools, and documentation aids
    Production data for testing
- Compliance with information classification standards
    Data classification requirements
    Critical systems analysis requirements
    Data backup and off-site requirements
    Records retention requirements
- Promotion to production requirements
    Proper authorizations required
    Peer review of coding
    Structured walk-through completed
    Documentation completed

APPLICATIONS DEVELOPMENT SAMPLE PROCEDURE

Installation Phase

Definition: Installation moves a system into production status.

Objective: The objectives of installation are to:
- Deliver the finished system product to the customer.
- Transfer operational responsibility to operations.

Discussion: Installation is required with all projects.

During installation, the project team converts the system's code from test to production status and converts the system's databases from their old structure to the new structure. The project team works closely with the customer, hardware or software vendors, the user support team, and the programming staff.

The project team's goal is to turn over responsibility for the system from the project team to the customer and the data center operations staff.

User procedures are issued to the customer during installation. In addition, customer satisfaction surveys are distributed so that customers have a formal means to give feedback on the new system.

The system run book, which contains the operator procedures, is turned over to the computer operations staff.

The following items are turned over to the product support staff, or whomever will support and maintain the system after installation:
- System profile
- Program profiles
- System architecture definition report
- System data model report
- Database definitions report
- Function specifications
- User procedures
- Training plan
- System test plan
- Installation plan

SYSTEMS PROGRAMMING SECTION

To the computer security professionals, the systems programmers are often viewed as the loose cannons of data processing. In many organizations, these employees are often allowed to ignore the organization's dress code. All too often, this kind of action sends the wrong message to the other employees. While you would not wish to make automatons of your employees, every employee must follow the rules established by the organization. While an employee's dress may appear to be a minor point in the overall control environment, it can convey a message from management that some employees are above the rules. Your job is to make sure that controls are applied uniformly across the organization, that they support the business function, and that they are followed by all departments.

Controlling systems programmers is often a difficult assignment. Some systems programmers are very territorial; they stake out their ground, both in the office and in the system and will defend what they perceive as their own. Some successful security programs have used the systems programmers' territoriality to elicit support for the controls required. The benefits gained by obtaining support from the systems programmers are worth whatever effort it takes. It is the creative work of the systems programmers that allows the data processing department to fulfill its objectives.

Systems programming overview statement

As changes to operating systems are implemented, effective controls should be maintained and changes should be documented so that only authorized and tested modifications are executed.

Systems programming minimum requirements

Usually, some of the systems programmers have come from applications programming and are accustomed to a structured environment. The controls required in the systems programming area are very similar to applications programming and can be quickly implemented. The following areas should be addressed:

- Authorization
  System changes authorized
  Authorization requirements for each system established
  Access to production system libraries restricted
- Promotion to production
  System changes thoroughly tested
  Structured walk-throughs prior to implementation
  Peer reviews conducted
  Back-out procedures developed prior to implementation
- Records management
  System maintenance logs retained
  Authorizations retained
  System listings retained
- Contingency planning requirements
  Systems stored off-site
  System documentation stored off-site

SYSTEMS PROGRAMMING SAMPLE PROCEDURE

Appropriate controls over technical support activities should be developed to ensure the stability and integrity of system software. Technical support management is responsible for ensuring compliance with the established procedures. The following are the minimum requirements necessary for controlling these activities:

1. All changes or new developments to the operating system, subsystem, or locally developed "technical programs" must be supported by documented authorizations.
2. Adequate pre- or post-implementation reviews must be conducted to ensure that only authorized changes are made.
3. Controls must be established to ensure that only authorized changes are applied.
4. Prior to system implementation, the proposed modification must be reviewed by the computer center management.
5. The customer must be informed of any changes and their impact.
6. Adequate backup procedures must be carried out prior to implementing the system modification.

DOCUMENTATION

Listings of system support activity with proper authorizations are to be retained by the local unit

TELECOMMUNICATIONS SECTION

With telecommunications extending all over the world, organizations are paying increasing attention to remote access of computer systems. For a compelling look at what can happen to an open network environment, take a look at Clifford Stoll's book, *The Cuckoo's Egg*.

For your telecommunications section, you should establish levels to control remote access to the system and transmission of data from the system. If you are transmitting financial data, your organization may be subject to federal regulatory guidelines. The procedures must protect telecommunications both for authorized employees and from unauthorized outsiders. When formulating a policy on telecommunications, you will also have to address the possibility that classified data may be transmitted in error to an unauthorized location.

Telecommunications overview statement

The ability to access the system from remote locations puts the organization at greater risk because of the possibility of unauthorized access to data and systems. An effective telecommunications security program will ensure that only authorized employees with a clear business need will be granted access and that eavesdropping, wiretapping, electronic snooping, or other interference is prevented or, at least, minimized.

Telecommunications minimum requirements

The elements of an effective telecommunications policy are actually a microcosm of an organization's total security program. Access to the system must be restricted to authorized users. Once in the system, users should only have access to the level of data for which they have been granted access. Additionally, software programs must be safe from unauthorized modification. Once in-house security is in place, you can concentrate on preventing break-in attempts by outsiders. The following items are key areas that should be addressed when creating telecommunications procedures:

- Software controls

    Updates to telecommunications software thoroughly tested

    Access to bulk data transmission files restricted

    Access to system parameter files restricted

- Operational controls

    System recovery/restart process controlled

- Hardware protection

    Communication closets and cabinets locked

- Transmission hardware

    Relay closets

    Telephone rooms

    Modems, control units, and other transmission devices

- Remote user identification

    Access control package userid and password

    Remote dial-up token password or dial-back controls

- Encryption requirements

    Key controls for encryption software

    Access control list current and monitored

TELECOMMUNICATIONS SAMPLE PROCEDURE

Dial-up Requirements

In order to reduce unauthorized access via the telephone system, there are four basic requirements that should be met. Typical mainframe and minicomputer operating systems, when properly used, may be able to take care of all or part of the problem; however, no unenhanced microcomputer operating system can do so. If these requirements are not adequately met by the host, then add-on hardware or software may be required to meet minimally acceptable standards.

    1. User identification and authorization. Access control packages such as CA-ACF2, RACF, CA-TOP SECRET, VMS, or locally-developed software are the primary methods of access control. When this capability is weak or nonexistent, or the system contains information of a classified nature, then an external hardware mechanism should be installed to augment this process.

2. Security event logging. All dial-up communication activity between host and user should be monitored in order to uncover intrusion attempts or successes.

3. Limiting the attacks. The system software should limit the number of repetitive user sign-on attempts per dial-up connection. If the software is inadequate, then alternative methods must be incorporated.

4. Encryption. If the information that is accessible via dial-up connection is classified secret or confidential or the system has been identified as critical, then additional protection from disclosure or tampering must be implemented. This form of protection should meet those requirements discussed in the classification matrix.

OPERATIONS SECTION

Operations security may be one of the longest sections in your security manual because operations involves so many employees: Operations schedulers, tape librarians, DASD managers, computer operators, and customer service employees. Each category of personnel requires a separate level of responsibility. With so many employees involved, the risk of exposure for the organization is great, and proper controls must therefore be established for all operations activities.

Operations overview statement

Computer operations controls the access to and processing of the organization's information resources. To ensure continuity of data processing capabilities, control measures should be established and enforced to reduce the risk of modification or disclosure of data and programs and reduce the risk of equipment failure or malfunctions because of unauthorized access to hardware and to program and data files.

Operations minimum requirements

In this section of your procedures manual, you should refer to the access controls established for the physical security of the computer center whenever appropriate. In addition to physical security, the operations security section should identify a daily operations manual to be used by each job area. This section in the manual should also outline the security measures that should be described in detail in the operations manual.

The following chart identifies the kinds of procedure manuals required for the data processing organization.

FIGURE 14 **DATA PROCESSING PROCEDURES**

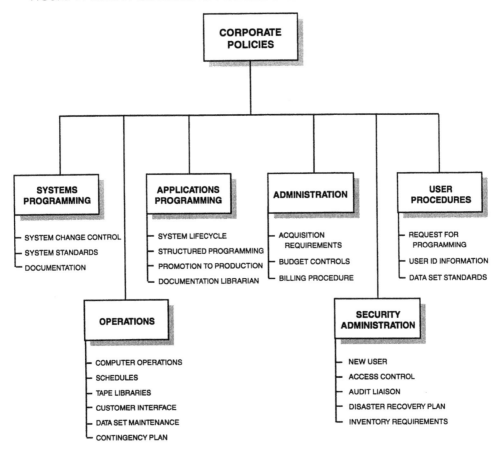

For the operations and administration staffs of data processing, the operations procedures should include a section on audit and inventory. Check the audit guide developed by the auditing department for suggestions on what to include. At a minimum, the following areas should be included in the operations security procedures:

• Inventory and control

   Vital physical records updated annually and stored off-site

   Annual review of electrical requirements

   Semi-annual inventory of tape library

   Annual inventory and testing of fire detection devices

   Annual inventory of DASD storage devices (not required with nonremovable DASD devices)

Annual review of special and multiple form output to ensure that they are still required

Quarterly inventory of card access system

Inventory of computer equipment (terminals, printers, and other hardware)

• Computer subsystem controls

Access control package (RACF, CA-ACF2, CA-TOP SECRET, etc.)

DASD management package (including archive requirements)

Scheduling package

Tape management system

• Physical access controls

Control of computer room areas

Control of remote processing areas (control closets, telephone room, cable rooms, and so forth)

Stock storage areas

Tape library and remote tape storage areas

• Emergency, backup, and disaster recovery plans documented and tested at least annually

• System and program ABEND process established

Program ABEND reporting and reconciliation

System incident reporting

Emergency system access controls

OPERATIONS SAMPLE PROCEDURE

Physical Access Controls

Physical access to areas containing data processing activities must be restricted to employees with a clear business need for access.

Access to critical areas may be controlled by:

1. Installing card access control systems.
2. Installing cipher lock controls.
3. Providing key-operated locks on doors.

A computer-based card access control is preferred, as it provides a high degree of protection as well as an audit trail on access to the area. The card access control system is recommended for securing the most critical computer activities.

In addition to controlling physical access, other security controls may need to be incorporated. These controls include the following:

1. Escorting visitors and documenting their presence.
2. Documenting after-hours access to computer areas.
3. Providing alarms or closed circuit television systems.
4. Providing security service after regular business hours.

5. Restricting access to computer operations, including tape storage areas, to authorized persons (generally those responsible for operations and maintenance).
6. Developing procedures for controlling keys to access doors, equipment with key locks, and lockable disk storage.
7. Installing key locks on all printers in uncontrolled areas that are connected to devices maintaining critical data.
8. Physically securing terminals if the security architecture is structured to allow access by physical device rather than by userid and authorization checks.
9. Providing visual observation of computer equipment during business hours if the equipment is not in a secured area.
10. Developing procedures to ensure that sensitive information is not stored on hard disk peripheral equipment unless an access control system is in place.

Access control personnel are to be notified whenever there is a change in employee access status (such as transfer, termination, or new assignment).

INFORMATION CLASSIFICATION SECTION

As a security program matures, the difference between confidential and sensitive data often becomes obscured. In an organization working under Department of Defense (DOD) contracts, there are clear-cut guidelines governing classification of data; however, in other work environments, the classifications often become confused.

Refer to Chapter 3 if you need more information about the controls necessary for an information classification system.

MICROCOMPUTER SECTION

Most organizations have witnessed an explosion of microcomputer use. In 1985, FBI Director William Webster noted that there were about 3.5 million personal computers in use throughout the business world, most of them in financial organizations. Less than three years later, *Business Week* indicated that there were over 33 million desktop computers, with their use no longer limited to business. At the Computer Security Institute Conference in June, 1990, a figure between 58 and 60 million personal computers was given. The increase in decentralized computer use has eroded many of the traditional security precautions of organizations.

Yet increasing decentralization merely underscores the need to establish controls and procedures. As computers move away from the regulations of the mainframe computer facility, there is a decided increase in the potential for unauthorized access and for modification, disclosure, and destruction of data. Clearly, in this environment, the ultimate responsibility for information protection lies not with the data processing department, but with the management of the user department and, ultimately, the organization's senior executives.

Microcomputer overview statement

Microcomputer security depends on the vulnerability of the equipment to theft and the sensitivity of the information stored in the equipment. Management is responsible for determining security requirements, instituting the necessary security precautions, and supporting the selected approach with a justifiable business rationale.

Microcomputers are defined as microsystems, personal computers, desktop computers, portable computers, laptop computers, and small business computers.

Microcomputer minimum requirements

As you define your policies and procedures, you should avoid using the term personal computer. Many employees take the term personal computer to mean just that: *their* computer. When they leave a department, many employees try to take the system with them. Additionally, since employees assume the computer is theirs, they feel that they can do whatever they want with it. In your written procedures, try to use a term such as desktop computer that does not have the same connotations as personal computer.

The microcomputer connected to a local area network has the same capacity that many computer centers used to have. Therefore, all the security elements required for a mainframe computer facility should apply to a department's microcomputer system. At a minimum, the following areas should be included in a section on microcomputer security:

- Maintaining an inventory of all units, including hardware and software.
- Developing requirements to secure access to the units during and after working hours.
- Maintaining a secure environment for all system control units, file servers, or master units.
- Maintaining lists of all supported hardware and software.
- Establishing a central ordering point.
- Establishing controls for removable disks.

- Establishing labeling requirements for disks.
- Publishing a software code of ethics.
- Creating department contingency plans for continued microcomputer use.
- Establishing data backup and off-site storage requirements.
- Establishing controls for storing sensitive data on hard drives.
- Establishing virus protection controls.
- Creating department documentation requirements for locally developed programs, macros, databases.
- Creating control mechanisms for the use of organization-provided portable microcomputers.
- Establishing employee training requirements.
- Assigning a PC coordinator for each functional work area.

MICROCOMPUTER SAMPLE PROCEDURE

Using Employee-Owned Hardware and Software, Shareware, or Evaluation Software

Employee-owned hardware and software, shareware, and evaluation software are not to be used in conducting company business without written management authorization. Managers have the responsibility to ensure virus contamination checks are performed on all employee-owned, shareware, and evaluation software. Virus detection software is available for use in testing for contamination. The PC security administrators are resources, available to the managers, to assist in this process.

It is the responsibility of the requesting employee to:

1. Obtain management authorization to use employee-owned hardware, software, shareware, or evaluation software within the business environment.
2. Inform the department's PC security administrator.
3. Evaluate for the presence of virus contamination before you use the software on organization-owned hardware.
4. Keep user documentation on-site while software is installed at the work environment.
5. Remove the employee-owned hardware or software from the work environment when complete. The original written authorization must be initialed and dated by both the manager and employee indicating removal from the location.
6. Complete the Personal Computer Checklist.
7. Retain the checklist for as long as the software is installed on an organization-owned personal computer and for 90 days after it is removed.

If you have any questions, please contact your manager.

FIGURE 15 **PC SECURITY CHECKLIST**

PC SECURITY CHECKLIST

This checklist is for use whenever employee-owned hardware, software, shareware, or evaluation software is recommended for use within the organization business environment.

Recommended by:_____Date:_____

Manager's signature:_____Date:_____

NAME of HARDWARE or SOFTWARE:_____

SOFTWARE TYPE   Employee-owned_____   Shareware_____   Evaluation_____

Reasons for use:_____

_____

SOFTWARE DOCUMENTATION (Reference manuals, user manuals or other documentation.)

Present at which location?   Yes_____   No_____
If not, why?_____

_____

VIRUS DETECTION

Date of virus scan:_____   Software or method used:_____

Operator initials:_____   Results:_____

INSTALLATION

Date installed:_____   Installer's initials:_____

REMOVAL

Proposed end date:_____   Actual end date:_____

Remover's initials:_____   Manager's initials:_____

SOFTWARE CODE OF ETHICS

Unauthorized duplication of computer software is contrary to our organization's standards of conduct. We disapprove of such copying and recognize the following principles as a basis for preventing its occurrence:

- We will neither commit nor tolerate the making or use of unauthorized software copies under any circumstances.
- We will provide legitimately acquired software to meet all standard software needs in a timely fashion and in sufficient quantities for all our computers.
- We will comply with all license or purchase terms regulating the use of any software we acquire or use.
- We will enforce strong internal controls to prevent the making or use of unauthorized software copies, including effective measures to verify compliance with these standards and appropriate disciplinary actions for violation of these standards.

ORGANIZATION	NAME AND TITLE
DATE	SIGNATURE

CONTINGENCY PLANNING SECTION

As identified throughout the volume contents, contingency planning must be woven into every area of the business cycle. What has been commonly called disaster recovery planning is now considered the data processing portion of an overall organizational emergency control plan. These emergency programs are also called business resumption plans, contingency operation plans, or emergency control plans. Whatever they are called in your organization, contingency planning is an organization-wide program and is normally under the control of the head of security.

In this discussion, the term disaster recovery plan will be used to identify the part of the organization's business resumption plan that addresses data processing. Data processing, while an important part, is obviously only one element of the organization's overall plan.

Contingency planning overview statement

Contingency planning is a business concern of the entire organization. All department managers must be responsible for the development of business resumption procedures within their own areas as well as for participation in the business resumption plan (BRP) for the entire organization.

DISASTER RECOVERY PLAN (DRP)

Computer and information technology and systems have developed rapidly in the past decade. The processing and delivery of information through improved technology has expanded management dependence on the availability and reliability of automated systems. Management is responsible for controlling this technology and its use in the organization. The continued availability of information systems is a joint effort of the data users and managers. For the purpose of clarification, the section of the BRP that addresses computer and/or information processing shall be known as a disaster recovery plan (DRP).

Disaster recovery plan minimum requirements

Disaster recovery planning generally consists of three phases. The initial phase, completed by the data processing staff and the users, identifies applications that are critical to the continuation of business. Once the critical applications have been identified, the organization can calculate loss outage or determine how long the organization can continue to function without the identified critical applications. Calculating loss outage will allow the organization to establish the priorities for a disaster recovery plan.

When the critical applications and systems have been identified, the second phase of the disaster recovery is to identify the hardware and software necessary to support the application. Based on the size and time requirements for recovery, the application or system owner will select a recovery strategy. The options for recovery include reciprocal agreements, cold sites, and hot sites. The choice of options should be based on the need and cost of the specific recovery process.

During the second phase, the actual plan will be developed. The first item in a disaster recovery plan is to name a coordinator. The coordinator will assemble the various teams and team leaders, including the following:

HUMAN SERVICES TEAM

Software team	Operations team
systems	library
applications	hardware
Communications team	Security team
data communications	data
communications software	physical
User teams	Database team

The final phase of disaster recovery is to test the plan on a regular basis. At least one complete test of the plan should be made so that all the critical applications go down and are recovered. A disaster recovery plan is a never-ending process; as soon as it is completed, it must be reviewed and updated. The worst plan is one that hasn't been reviewed for six months. Like policies and procedures, writing the disaster recovery plan is the easiest part—testing and updating are the true challenges.

A number of relational database packages currently are used to automate the disaster recovery planning process. One such product is the ChiCor Total

Recovery Planning System (TRPS). This package allows the timely update of the teams required for the business resumption process and may be useful for your organization.

The minimum requirements for the data processing DRP include:

- A DRP administrator should be designated for each unit. A unit would include any subprocessing work area. For example, you may have a mainframe computer facility and then some remote processing areas or even another data center located away from the main center. Wherever there is a concentration of computer processing capacity, a DRP coordinator should be assigned. Microcomputer locations should also be included in this rule, although to a lesser extent.
- Top management should review and approve the initial DRP and subsequent annual updates. In the financial industry, updating is a federal requirement.
- Information owners are responsible for notifying the data processing staff about the information to backup and the frequency required.
- The data processing staff is responsible for the timely backup of the system files and off-site storage of these files.
- Hard copy documents must be copied and stored off-site.
- Requirements for testing at least a part of the DRP must be established.
- The DRP must be coordinated with local security and law enforcement personnel.

DISASTER RECOVERY PLAN SAMPLE PROCEDURE

The Goals of DRP Test Planning

When preparing to test a DRP, either completely or in part, it is important to specify particular test objectives for each of the areas that are to be evaluated. The objectives can differ for the areas being tested and the type of test being conducted. There are, however, some general objectives that apply to any type of contingency testing:

- Verification of the completeness and precision of the DRP information.
- Evaluation of the performance of the personnel involved in the exercise.
- Appraisal of the training and awareness of noncontingency team members.
- Evaluation of the coordination between the contingency team and external vendors and supplies.
- Measurement of the ability and capacity of the backup site to perform prescribed processing.
- Assessment of the vital records retrieval capability.
- Evaluation of the state and quantity of the equipment and supplies that have been relocated to the recovery site.
- Measurement of the overall performance of the operational and data processing activities related to maintaining the business entity.

FIGURE 16 **RECORDS MANAGEMENT LIFE CYCLE**

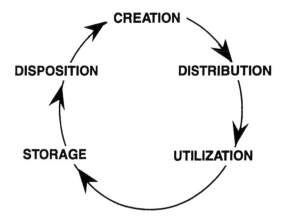

RECORDS MANAGEMENT SECTION

Records management is the systematic control of the life cycle of information, as shown in Figure 16.

A number of organizations refer to records management as records retention. Records retention is a bit misleading, however, because retaining records is only a part of a records program. Records must also be systematically destroyed at the end of their usefulness to the organization.

Records retention is also commonly considered synonymous with disaster recovery planning, but is actually distinct. Vital records are required for emergency control plans and disaster recovery plans, but other records are kept to fulfill government regulations and satisfy the organization's historical needs.

Records management overview statement

As an asset of the organization, information is at least as valuable as cash reserves. In order to use information, a structured information management program provides the strategy for planning, organizing, and controlling information for the organization. Records are retained on a consistent basis to meet federal, state, and local regulations. Records are also retained to satisfy the operational needs of the organization, such as the information that is required by local units to complete assigned tasks or information that is historical in nature. If the value of the information diminishes over time, the information is disposed of using the guidelines established to minimize risks and liabilities for the organization. All information, regardless of the media (electronic, hard copy, microfilm, and so forth), is subject to retention policies and includes, but is not limited to, financial, engineering, manufacturing, personnel, legal, data processing, and other departments.

Records management minimum requirements

As discussed, records management must be integrated into the policies and procedures of the entire organization, and the systematic identification, retention, and destruction of records must be adopted by all employees in their daily operations. At a minimum, the records management section must establish controls for:

- Types of information

  Organization assets

  Information on the stockholders

  Employee records

  Tax records

  Other essential records

- Procedures

  Information required for compliance

  Retention period for each record type

  Location of records

  Creation of records

  Personnel responsible for records storage

  Review and/or destruction of documents

RECORDS MANAGEMENT SAMPLE PROCEDURE

Records Retention and Management

Information is a corporate asset. In order to utilize this asset, a structured information management program constitutes a major strategy for planning, organizing, and controlling information for the corporation. Records are retained on a consistent basis to meet government regulatory requirements and fulfill operational needs. Regulatory requirements for retention are established by federal, state, and local statutes. The operational needs of the corporation can be defined as information that is historical in nature and required by local units to complete assigned tasks and that diminishes over time in usefulness. Information disposed of during the normal course of business by adherence to established retention guidelines minimizes risks and liabilities for the corporation. All corporate information, regardless of media (electronic, hard copy, microform, etc.), is subject to retention policies and includes, but is not limited to, financial, engineering, manufacturing, personnel, legal, data processing, etc.

CHAPTER REVIEW

1. Management is responsible for creating and implementing all administrative policies, procedures, practices, standards, and guidelines.
2. The control type are:
    Management
    Preventive
    Detective
    Containment
    Recovery
3. All controls must be based on a cost-benefit analysis.
4. Physical security requirements:
    Employee identification process
    Physical plant controls
    Fire control
    Inventory of equipment
    Electrical requirements
    Housekeeping requirements
    Vital physical records storage
    Guidelines for records disposal
5. Security administration minimum requirements:
    Access control package administration
    Employee userid and password administration
    System access terminal requirements
    Log monitoring
    System maintenance requirements
    Contingency planning requirements
    Records management requirements
6. Application development minimum requirements:
    Request for service
    Programming standards methods
    Information classification requirements
    Promotion to production controls
7. System programming minimum requirements:
    Maintenance authorization
    Promotion to production
    Records management
    Contingency planning requirements
8. Telecommunications minimum requirements:
    Software controls

Operational controls
Hardware protection
Transmission hardware
Remote user identification
Encryption requirements

9. Microcomputer minimum requirements:
Authorized usage controls
Right to audit controls
Acquisition requirements
Basic equipment requirements
Software code of ethics
Virus control
Training requirements

10. Disaster recovery planning minimum requirements:
Critical applications definition
System backup requirements
Off-site storage controls
Disaster plan documentation
Coordination with overall business resumption plan

11. Records management minimum requirements:
Organization assets
Stockholder information
Employee records
Tax records
Compliance requirements
Retention period for each record
Location of record
Creation of record

Exercise 1

Using the material just discussed, decide where each of the following topics should be placed in your procedures manual:

1. Establishing a charge-back system for a software development project.
2. Implementing project management techniques.
3. Developing background verification checks for new hires and contract employees.
4. Establishing a clear termination policy and procedure for employees. Would your policy apply to contract employees?
5. Controlling the acquisition of hardware and software. Would your policy cover PC hardware and software?

6. Controlling sensitive forms and negotiable documents (i.e., checks, purchase orders, vouchers).

Exercise 2

Taking into consideration all the security requirements discussed in this chapter, develop an effective terminal sign-on screen.

CHAPTER 5

Review Panel

As you develop policies and procedures, you will need to have your work reviewed by someone from your organization with a basic knowledge of your topic and some interest in seeing the project completed. The writing process often drives the writer into a closed world. If you have been spending time gathering sources of information (see the Appendix for further information), your focus may be rather limited, and you may not realize the precise controls that should be implemented for your organization. So, from time to time, you will need to check your perspective and subject your work in progress to a fresh point of view.

Computer security personnel must always remember to use due, deliberate speed in implementing programs. All too often, after attending a security conference or spending time researching topics, the practitioner goes overboard and attempts to implement all the new controls acquired from the conference or research. It's important to recognize that what you see at a conference or read in a published text is the accumulation of years of experience. Implementing all of the known computer controls at one time could send the employees of your organization into security shock. Therefore, your research should be geared to the level of security suitable for your organization, and your conclusions should be reviewed regularly to ensure a proper balance between the ideal security environment and the reality of your organization.

In the early development stage, a few trusted individuals from each affected area should be asked to contribute their ideas on your procedures. Alternatively, the same individuals could be asked to draft sections of the procedures. After all, the employees who work in a department obviously know how it operates. By soliciting their input, you garner support for the finished product.

Once the draft is complete, you should establish a panel of employees to review the document. The panel will provide an objective review of the draft contents and will establish a base to gain acceptance for the manual throughout the organization.

Before sending out material for review, you should meet with your organization's policies and procedures department to ensure that the draft resembles existing organizational policy. This may sound like a minor item. However, management has grown accustomed to the layout and wording of existing documents. If your material is different, regardless of the content, it may be dismissed before it can be reviewed.

The review process generally takes three rounds. The initial review should include a small group of interested individuals. For many organizations, a good place to start is with the data processing steering committee. This group is usually established to oversee and manage new systems implementation, hardware purchases, and funding and appropriation approvals. An additional responsibility should be to review and comment on computer security policies and procedures. The group should include representatives of key user departments and senior data processing management.

During the initial review, you should consider providing the hard copy for any referenced material. The review process gets bogged down when a reviewer is forced to search for letters, procedures, laws, and other documents that are referenced in the manual. You should keep distractions to a minimum when reviewers are checking the draft document. All too often, if the reviewer is distracted, the result will be a negative review.

As a courtesy, you should inform the reviewer that the document is coming prior to distributing the draft. Sending procedures to an individual without notice can cause a great deal of needless grief. If you first send the document to the data processing steering committee, try to get on the agenda for a committee meeting so you can give a short presentation of your objectives. The material developed in the mission statement exercise would be very appropriate here. The following guidelines apply whenever you send something out for review:

1. Inform the recipient that the material will be coming.
2. Check to see that the recipient receives the material.
3. Follow up to see that the material is being reviewed.
4. Ensure that comments are returned in a timely manner.

If you come up against a roadblock or find that some of the reviewers are slow to respond, you can use a steering committee meeting as a forum to discuss the current status of the procedures. Never identify by name any individuals who are delinquent. You may, however, make references to the number of copies that have been mailed and the number that have been returned; this is the tactful way to make your point. Additionally, you can indicate that if responses are not returned by a specific date, it will be assumed that there are no comments. While this is a good ploy, you should still call the nonrespondents to verify that they did not wish to respond and to prevent any ruffled feathers. These and other political issues involved with selling the manual to management will be discussed at greater length in Chapter 6.

ACCEPTING COMMENTS

Whenever possible, the comments provided by review panels should be incorporated. If there is a conflict between returned comments and/or the comments and the proposed procedure, then you must rank the comments for their priority. The following scale lists rankings from highest to lowest:

Legal staff

Management (yours is always first)

Audit staff

Physical security

Personnel

Labor relations

Data processing

Key units

Selected user departments

When all the comments have been considered and the draft has been updated, a second round of reviews should be started. During this cycle, you should expand the number of reviewers to include at least one representative from each of the ranked areas. During this cycle, you should also send review copies to any overseas operations, affiliated units, subsidiaries, and the like. A cover letter should be attached that briefly describes the reason for the review process. You should mention the fact that this is a second-round review and set a response deadline, generally a month from the distribution date. The following is a sample cover letter:

SAMPLE COVER LETTER

Date: January 31, 1991

Subject: Information Security Procedure Manual
Proposed Update Draft

To: W. Anders - Legal Staff
J. Acback - German Operations
R. Buddy - Canadian Operations
L. Degg - Personnel
G. Dickerson - Research Laboratories
J. Diet - Purchasing Policies & Procedures
Directors - Audit Staff
J. Elwood - Personnel Policy Staff
E. Flash - Financial Staff

G. Guthery - UK Operations
R. Hardy - Comptroller
T. Jefferson - Engineering Operations
C. Knoll - Mexican Operations
G. Kettler - Security
D. Guard - Pacific Rim Operations
W. Lithgow - Data Processing
M. Metzner - Divisional Finance Group
O. Marshall - Aerospace Division
J. Rush - Chief Legal Counsel
F. Scheen - Internal Controls
B. Striker - Design Services Operations
A. Townsend - EDP Audit
R. Tracker - Data Processing Investigations
P. Turner - Electronic Development
J. Wright - Hourly Personnel
J. Venezia - Manufacturing Group

The attached document has incorporated comments from the initial set of proposed updates sent out to the data processing steering committee on October 24, 1990. Because of your expressed interest in the concept of improved information security, you are being provided with the opportunity to review the proposed manual and to present your comments to the manual coordinator. To facilitate your review, an appendix containing all referenced corporate procedures, policy letters, and other documents has been provided. It would be appreciated if your comments could be returned to the manual coordinator by February 28, 1991.

If you have any questions, please contact the manual coordinator at 555-8705.

The letter should be signed by the vice president or director of your department. Copies of the letter and manual should be sent to all senior managers with a vested interest in completing the project.

After the second distribution of the draft manual, your work will begin in earnest. Generally, the individuals to whom you have sent the document will pass the material, unread, to a subordinate for review and comment. The draft will then be dissected and examined to determine how much additional work will be caused by its adoption. The individuals reviewing the manual at this level are not interested in any altruistic goals; they already have too much to do and are not looking to add to their workload. You will therefore have to establish a dialogue with the reviewers to answer questions, provide insight, and listen to alternatives. Your task during this phase is to oversee the manual and keep the review process moving.

During the second review process, you may find it beneficial to have a meeting with the reviewers and allow them to discuss their concerns in an open forum. At any meeting of this type, you should include the individuals you selected to review or write procedures during the development phase. These individuals can provide useful perspective on the manual.

Do not be surprised if some of the reviewers ask for extensions. You should be flexible about delays, yet also keep in mind your own deadlines and timetables. Keep after the reviewers and call them at various times during the review period to keep the process moving. If there are comments that cannot be implemented, then you will have to contract the submitting department and explain why the comment was not accepted.

When you finally receive the last of the comments, you will be ready for the final round of the review process. This time your cover letter should notify the reviewers that this is their final chance to review the proposed manual before it is published. Some reviewers need to see the word final before they send any comments. The updated document should be sent to the same list of people as in the previous round. Your material should indicate what comments were added to the procedures. This time the review period should be a maximum of two or three weeks.

REVIEW OF SECURITY POLICIES AND PROCEDURES PROCESS

1. The first step in establishing a computer security program is to formalize an overall information security policy—which is accepted and approved by management—recognizing the need for information security. The policy should outline the security responsibilities throughout the organization. This will provide the infrastructure from which all other security procedures, practices, and standards will be created.

2. Develop a management-approved mission statement that establishes the role of the information security officer and the responsibilities associated with this position.

3. Develop a written policy on the classification of data according to its risk of exposure and loss. Include the requirement for regular review of the classification levels so the classification levels can be downgraded as appropriate. The policy should also address the issues of integrity and availability of the data and system.

4. Establish a written policy on the ownership of information, transactions, and assets, and clearly describe the responsibility of the owners, users, and custodians.

5. Create a written policy on records management, including standards on length of retention, type of storage, and destruction procedures.

6. Develop a written policy on updating risk assessments.

7. Establish the audit function and its roles of internal consultant and

compliance checking. A right-to-audit clause must be established to inform the employees on regulations regarding their own personal data. The audit right must be extended to all data files (including those on disks, cartridges, tapes, DASD, etc.) and all data processing resources (including mainframes, minis, micros, electronic mail, voice mail, etc.).

8. Develop written promotion-to-production procedures for all new or updated applications, transactions, systems, etc. Include the need for peer review of source code and use-testing prior to implementation.

9. Create written standards for creating and testing data processing disaster recovery plan. Include requirements for the microcomputer community.

10. Generate written procedures for implementing and reconciling a regular inventory of all assets, both physical and intellectual.

CHAPTER REVIEW

1. Establish a small, knowledgeable initial review panel.

2. Do not create all the procedures by yourself. Seek out personnel in areas affected by the controls and gain their expertise and assistance in this process.

3. Be certain that the procedures resemble the procedures currently being used in your organization.

4. Try to get on the agenda of the data processing steering committee to present your program and solicit the committee's support.

5. Wherever possible, accept and implement the comments created by the reviewers. At the very least, contact the reviewer and explain why the comment cannot be included.

6. If there appears to be a conflict, set up a meeting, at the respondent's location if possible, to resolve the problem.

7. Be persistent. You are going to have to keep after the reviewers to get their responses.

Exercise # 1

Using the information just discussed, develop a list of the areas that should be included as part of the review process.

Include areas for the:

Initial review

Second review

Final review

Exercise #2

List the members of your final review panel and rank them according to the priority you will give to their comments.

Exercise #3

Draft the cover letter for your initial review draft.

CHAPTER 6

Selling the Program

SECURITY VERSUS REALITY

Security is more than an access control package, programs, locks, or passwords. Take, for example, a typical computer room facility. Most computer rooms have been divided into controlled zones and intense control zones. Each zone is protected by an access control card reader, so that an authorized employee would have to pass through at least two card readers to gain access to the computer room proper. To be in compliance with the fire code, the computer room has emergency fire exits with alarms along the outside walls. Take a walk through the computer room late at night. You will often find that the emergency exit doors that are closest to the bathrooms have the alarm taped shut and the doors propped open. Security control measures work only when the employees are properly motivated to comply with them.

TARGET GROUPS

There are three groups to which you have to sell the information security program. Each group has its own set of values and areas of concern. The three groups are:
- Management
- Data processing
- Users

To gain acceptance for the security program, you should start with top management and then approach the data processing and general users.

APPROACHING MANAGEMENT

The entire program of computer and information security will fail unless it is accepted and communicated by senior management. Managers must therefore be aware that they are required either legally or contractually to protect the organization's assets. Management's failure to provide adequate controls could lead to charges of imprudence. Your job is to educate managers about how they can fulfill the goals of security. Discussions about legal ramifications will usually only get the managers'

FIGURE 17 AUDIT COMMENT MATRIX

File Number	Location of audit	No Comment Report	Access to Confidential Data	Program Change Control	System Change Control	Access Controls IMS/CICS	ACF2 Controls	Physical Security/Fire Protection	Disaster Recovery Planning	Password Change Control	APF Access Control	Userid Security	Dial-up Controls
91-001	Banking Division		X		X				X				
91-002	Insurance Division		X	X	X	X	X	X	X	X			X
91-003	Products Division		X	X								X	
91-004	Computer Center—Eastern				X		X	X	X				
91-005	Canadian Operations		X	X			X	X			X		X
91-006	Mortgage Division	YES											
91-007	Personnel Administration			X	X			X	X	X			
91-008	Distribution Division			X		X	X		X		X		X
91-009	Products Division		X			X	X					X	
91-010	Japanese Operations		X		X				X			X	
91-011	South American Operations				X	X		X					
91-012	Electronics Division					X		X			X		
91-013	Accounting Center	YES											
91-014	Telecommunications		X			X			X				X
91-015	Transportation Division			X			X	X					
91-016	European Operations		X		X	X			X	X			X
91-017	Computer Center—Western			X			X	X					
91-018	Marine Division			X			X	X			X	X	
91-019	Employee Benefits Section		X	X		X	X		X				
91-020	Applications Development		X		X	X		X	X		X		
91-021	Manufacturing Division		X	X			X	X		X	X	X	X

attention for a short period. To gain their continuous support, you will have to appeal to them on their own terms.

Management is most likely to adopt your procedures when you provide a clear evaluation of the current level of security within your own organization. One way to evaluate the existing security policy is to study recent audit reports. You should include both internal and external audit staffs' reports for the past five years and use them to create an audit component matrix (see Figure 17). By establishing a pattern of security concerns, you will be able to pinpoint the weakest areas within your organization, and you can use this information as the basis for building your security program.

A walk-about is another effective way to evaluate security procedures. The walk-about should be conducted by you and someone else from the security staff during off hours (such as nights, weekends, or holidays). The aim of the walk-about is to quantify the level of information and computer security practiced by your organization. As an example, you might select a physical area (such as the second floor) or a logical area (such as purchasing) for your walk-about and perform the following tasks:

1. Take a count of the number of work areas (desks, cubicles, offices, and other work stations).
2. Determine the quantity and types of hardware (terminals, PCs, printers, modems, and other devices).
3. Establish the number and types of access and egress points.

Once these basic items have been noted, you should check to see if the following are secure:

- Offices
- Desks and cabinets
- PCs
- Disks
- Organizational data

During this review, you should check the terminals for any posted passwords: Take a minute to scan those self-adhesive notes and check the underside of the keyboards. This kind of activity may seem silly until you discover your first password and realize how common it is.

After you have compiled figures on security violations, you should prepare a report to management that identifies the area reviewed, the types of concerns you encountered, and the percentage of unsecured items. When preparing this report, you should simply describe the existing security conditions and avoid being judgmental or critical. Remember that, up until now, the employees have only received appraisals on how well and efficiently they performed their jobs. Securing the workplace did not affect the employees' job evaluations. By keeping your report objective, you can avoid a negative reaction from the departments you visited.

Another way to evaluate existing security procedures is to meet with your security staff and review the stolen property reports. A three- to five-year analysis will provide a theft pattern. Generally, without an effective security program, losses will increase annually, both in number and total value.

Studying audit reports, conducting walk-abouts, and determining loss patterns will assist you in making a clear evaluation of the current level of security in your organization. Addressing the problems in the existing security program should then provide you with a strategic plan and direction.

STARTING YOUR SALES PRESENTATION

Before you begin, you must decide what management expects of you. All too often people charge off on a project without a clear understanding of what they are supposed to achieve. Your charter or mission statement (as discussed in detail in Chapter 2) should give you the proper focus for your project.

As you prepare to approach management, it is a good idea to pick out some security issue that can be solved in a short period of time. Your goal here is to build credibility for the program. Security is a new area and your past successes mean little in your current role at your organization.

You should never go to management with a plan and a budget request to go from zero security to full compliance in one step. A radical change and price tag will definitely cause management to choke. Bringing your organization into a relatively secure processing environment is an enormous job. So limit the scope of your project and approach it in phases. Writing a set of policies, like the ones discussed in Chapter 4, is a low-cost way to begin.

Keep management informed

When writing policies and procedures, there is often a period of time when there appears to be no tangible results. You may find yourself deeply involved in research or in interviewing departments, reading a related book, or attending policy-writing workshops. During such times, you must be certain that management is kept aware of your progress and knows that this initial background work was factored into the overall plan.

A total security program requires an effective game plan that establishes the phases, timetable, and estimated cost of implementing the project. At each phase, you should prepare a communications vehicle to keep management and the users informed on the objectives of the current phase, the steps that are to be taken next, and the long-term objectives of the program. As you complete each project or phase of a project, management should be updated on your progress.

EXPECTED RESULTS

There are three basic tiers of management and each tier will respond best to a different selling technique. The next sections will examine each of the management levels and determine the best way to get and keep management support.

Senior Management

Senior managers are the bottom-line types; they usually want to know how their organization stacks up against the other organizations on security issues. The audit matrix, walk-about findings, and theft reports discussed earlier in this chapter can all be used to prepare presentation material for senior managers. As you develop presentation materials, be sure to compare the results of your findings with other, similar organizations. To develop these comparisons, you should establish liaisons with fellow security practitioners through your local computer security organization special interest group or your local chapter of the Electronic Data Processing Auditors Association (EDPAA). Information on professional associations is listed in the Appendix. By making comparisons with similar organizations, your management will be better informed when making their decisions.

Another method of comparing your organization with other organizations on security issues is through formal risk analysis. A typical risk analysis program identifies assets, threats, and probability of exposure (for additional risk analysis material, contact Computer Security Institute for their seminar on Practical Methods of Risk Analysis for Computer and Information Security). Once the three elements have been identified, various risk analysis programs then perform a series of calculations that allow you to quantify the risk level in your organization.

Senior managers have very limited time, so your best approach with them is to prepare a short, concise presentation and an in-depth handout. Normally your presentation should be about fifteen minutes long. In that time, you must quickly identify the problem, propose your solutions, and seek management's support. If the senior managers support your program, they will respond with public approval of it and will provide the funding to continue the work. To maximize your time with senior management, your presentation should include:

1. Statement of the problem. Briefly describe current conditions within your organization.
2. Cost impact of existing program. Analyze the cost to the organization if it does not accept the new security program.
3. Service impact of existing program. Describe the services that will be affected if security conditions are not improved.
4. Regulatory considerations. Discuss the government regulations or court rulings that will affect the organization.
5. Recommended solutions. Outline what steps must be taken to correct the situation.
6. Cost of solutions. Using the best data available, estimate the expenses for this program. Remember to compare the cost of the solutions to the cost of keeping the status quo (developed in item 2).

Middle management

Even though you have gained senior management's support, you should not assume that the rest of the organization will endorse your security program. Most managers (and employees) have been around long enough to know that, if they ignore or postpone implementing management directives, the changes often go away. Your goal with middle managers is to demonstrate how security will help them improve their job performance and productivity.

You must convince middle management that computer security controls are not counterproductive, but are actually a value-added activity. Change usually causes the most complaints in the workplace. Whenever some new procedure or control is introduced, employee efficiency slows down until the change either becomes part of the normal work function or is discarded as unnecessary overhead.

For this reason, you should implement your policies over a period of time. This does not mean that you cannot publish the procedures at one time. It does mean, however, that a timetable of compliance should be established. Your procedures should be presented to middle management as standard operating procedures for the organization. Just as a set of procedures has been developed to help new employees quickly adjust to their new jobs, your security procedures will help all employees comply with security requirements.

The best way to get your message to middle management is through presentations at staff meetings or some similar gathering. The presentation for middle managers can be a little longer than that for senior managers, about twenty to thirty minutes. You should briefly discuss senior management's support for the program, but most of your time should be spent on detailing how the new security policies will improve working conditions. For middle management, support will come when they actively participate in the program and see the positive results of your policies.

Lower-level Management

Lower-level managers are generally the first-line supervisors and managers. Their support is best gained from published, approved policies and procedures that are supported by upper management. With this group, your goal is to communicate the message through presentations, videos, newsletters, or other media. The general awareness program developed in Chapter 7 will discuss how to gain further acceptance for security procedures. Figure 18 gives a summary of management support expectations.

ACCEPTANCE

Before an organization can accept a security program, all the members of the organization must first understand why the program is necessary and what they will gain from implementing it.

Most employees understand or accept the need for structure in the workplace. There are generally rules established for starting times, work hours, job performance, dress code, vacations and so forth, and each job assignment has its own structure—including the tasks to be completed and how the department or organization expects these activities to be completed. Annual employee appraisals monitor the level of performance with the established job goals. The internal audit staff performs the same function for each work unit as the external audit company does for the organization as a whole. During each of these review processes, the established control structure is reviewed and modi-

FIGURE 18 MANAGEMENT SUPPORT EXPECTATIONS

GROUP	BEST TECHNIQUES	BEST APPROACH	EXPECTED RESULTS
SENIOR MANAGEMENT	Cost justification Industry comparison Audit report Risk analysis	Presentation Video Violations routing	Funding Support
MIDDLE MANAGEMENT	Demonstrate job performance benefits Perform security reviews	Presentation Circulate news articles Video	Support Resource help Adherence
LOWER-LEVEL MANAGEMENT/ EMPLOYEES	Sign responsibility statements Policies and procedures	Presentation Newsletters Video	Adherence Support

fied as required. Your goal in gaining acceptance for the new policies and procedures is to fit into the existing control structure.

Senior managers understand the need for security, but often do not understand how the organization may be exposed. Some of these managers simply lack exposure to computers. As the year 2000 approaches, computer literacy will be less of an issue. For now, however, you will probably have to educate management on the changes in computer technology. Most managers have little appreciation for the massive decentralization of computer activities, and many managers would have a hard time telling the difference between a 3270 terminal and a microcomputer. The level of computer sophistication at the senior management level may just not be there, because these are managers who did not use computers on their way up and now consider it foreign territory.

The data processing managers and personnel, on the other hand, usually understand the security weaknesses in the systems and applications. The data processing users access the system daily and have been active in developing and installing the system, so they know what the hardware and software can do and how to make it work efficiently for them. Because the data processing personnel know how the system functions, control procedures, such as the separation of duties discussed in Chapter 3, are especially important and should not be overlooked by the security staff.

Users also understand how to use computers to get their jobs done. Users generally have no desire to add any control mechanisms because they feel that additional controls will slow down their progress in getting their jobs done. Changing passwords, access controls, individual userids, and terminal time-outs are all considered intrusions by users who are mainly interested in fulfilling their job responsibilities.

When you approach these different groups within your organization, you will have to understand that all groups of people have different perceptions about security. Each group may understand a different component of the problem, but few put it all together. It is hard to get the big picture when you are faced with deadlines and production schedules. Therefore, while no program can be all things to all people, the security program must identify the need for security and the benefit of it for each of these groups so that they will not just accept, but endorse, the program.

Employee acceptance levels normally fall into three categories. About 10 percent of the employees will accept a new security program when introduced and endorse it as their own. About 10 percent of the employees are initially opposed to a new security program. The remaining 80 percent will need to be convinced that the security program is valuable, and they will be the focus of your selling program. The first group is already sold, so you do not need to concentrate on them. Often the group that is opposed will make the biggest noise, but most of them will end up complying with the regulations. Nevertheless, about 2 percent of this group consists of malcontents who complain about everything, even a raise (when they get a raise, they complain about how long it took and how small it is). The true malcontents must be isolated and controlled or otherwise they will wreak havoc on your security program.

As discussed, the users, managers, and data processing staff will not accept security measures with open arms. The makeup of your organization has a lot to do with how long this process will take. For instance, if you have an open shop (such as a university or research facility), implementing a security program will be far more difficult than if you already have a secure environment (such as a military contractor or a DOD agency).

When you get ready to sell the security program to your organization, you will have to take a step back and direct your acceptance presentations to the 80 percent of the workforce that can be persuaded of the program's value. Before employees endorse any new computer or information security policies and procedures, they must first be made aware of their responsibilities, and discussing the reasons why security fails is often necessary for employees to comprehend their responsibilities in the security program. The following examples show typical security problems in a variety of companies.

UNCONTROLLED OR INADEQUATELY CONTROLLED ACCESS

An employee in a midwestern facility was passed over for a promotion and decided to access the human resources management system to see who was better qualified than he was for the job. The employee accessed the data that lists employees by job classification level and then ranks the employees in each category from top to bottom based on previous appraisals. The employee felt this information was interesting and printed it out and made copies. He posted the copies in the cafeteria, by coffee machines, and on bulletin boards throughout the building. Publicizing this private information obviously disconcerted many employees and caused needless problems for the company. The breach occurred because the personnel director had never changed his default password to the human resources management system.

Each director or manager is responsible for protecting the information in his or her department. It is the responsibility of the manager to ensure that the information is properly protected. The personnel director is responsible for ensuring that employee information is protected from unauthorized access or disclosure. If confidential information about an employee is disclosed, the personnel director can be held civilly liable for failing to protect the informational assets of the company. Similarly, an engineer who is developing a new product has the responsibility to protect the integrity of the information used in development. If the test results are modified or altered and the company ends up creating a faulty product, the engineer can be held civilly liable for not protecting the information.

In another case, a Canadian firm was showing its new computer-aided design package at an architects' convention. The package was an expert system for designing building facades. The owner of the company demonstrated at one station, while his son was at another. After finishing a demonstration, the son turned his back on the computer for a few seconds to ask his father a few questions. When the son returned to his computer, the program disk was gone. Even worse, the stolen disk contained the source code. The company estimated the market value of the software package at between five and ten million dollars and said it represented twelve years of research and development.

VAGUE OR INADEQUATELY DEFINED RESPONSIBILITIES

A Wayne State University graduate student had her doctoral thesis on the hard drive of the IBM XT she kept in her apartment. Someone broke into her apartment and among the items stolen was her PC. Since she failed to make any backups, she was forced to recreate the thesis using the printouts left in her ransacked apartment.

In another case, a user in an engineering facility who left a PC on at lunchtime returned to find the hard drive reformatted. An investigation determined that that reformatting was probably an accident. It was also determined that the PC that was reformatted two days later was probably done in retaliation for the first one. Because accidents can happen, you should encourage your employees to secure their work area whenever they leave and especially when they leave for an extended period of time.

A corporation in Georgia had a rather lax backup and storage policy. Forty diskettes containing valuable information and ten more containing crucially important data were stolen. The ten diskettes were so critical that the corporate officers couldn't put a monetary value on the diskettes because, if they didn't get the diskettes back, they were out of business. The corporate officers were so desperate to regain the diskettes that they went to the local newspaper and took out a ransom ad. The company offered to pay to get the diskettes back—no questions asked. Fortunately for the company, the police recovered the diskettes and the ad never ran. It seems that a custodian came through the office area to clean it and found the diskettes sitting out. Reasoning that the diskettes were not valuable (otherwise they would not be left out), he took the diskettes home and was going to reformat them so his kids could play games on their new home computer.

In another instance, this time in a San Francisco firm, all of the accounts receivable diskettes disappeared one weekend and were never found. It took weeks to reconstruct the data using hard copy. Additionally, cash flow became a problem because the company had difficulty collecting bills.

The next example shows what can happen when an employee is not aware of his responsibilities. A drafting supervisor was sent to a class to learn how to use a PC database. Being a loyal employee, the drafting supervisor began to apply what he had learned when he returned to work. He sat down and created a database consisting of each employee's name, social security number, home address, home phone number, last appraisal rating, current salary structure, and so forth. The supervisor did not want to leave the database on his PC hard drive because he knew that other people had access to the PC, and he did not want to leave the diskette in his desk because other people had access to his desk. So he decided to hide the diskette under his desk and attach it with magnets. As you know, diskettes and magnets do not mix, and all of the data was scrambled. He did succeed in making sure no one had access to the data, but in the process, he eliminated access for himself.

To avoid repeating the supervisor's costly mistake, always store diskettes properly. Diskettes and magnets are not compatible. Placing diskettes next to telephones, electric pencil sharpeners, electric staplers, or anything with an electric motor can cause loss of data. Placing a box of diskettes on top of the heating system (such as a register or radiator) can damage the diskettes. Writing on a floppy diskette with a ballpoint pen or pencil can also damage it and may make it unusable.

If you live in New York City, San Francisco, Boston, or Washington, D.C., you should not put your briefcase on the floor of a subway. The floors on some trains contain magnetic fields from the system's power supply that are strong enough to erase the data stored on diskettes.

INADEQUATE TRAINING OF PERSONNEL

A financial department in the Midwest had 100 to 150 diskettes disappear one weekend. Some of the diskettes contained two to three years worth of financial data—and there were no backups. It took the department six months to rebuild the database using hard copy.

In another example, many departments were using a mainframe package called ADRS (A Department Reporting System) to develop information used within the department. When support for ADRS was dropped and the departments were given PCs with Lotus 1-2-3, there was a breakdown in the communication of responsibilities. After a few months, the secretary went to the operations department to have her Lotus files restored, as she had routinely done whenever she had problems with ADRS files. When informed that it was now her responsibility to backup data, she became very upset. She had to reconstruct the data using hard copy because she had never been told she was responsible for backing up the data.

Employees who do not sign off when completing their work or lend someone their password cause a serious breach of security. Just as each employee is unique, the userid to access the computer systems is unique. Each userid has a specific level of authority. If a password is lent out or employees do not sign off after completing their work, they are effectively giving a stranger a signed blank check.

Illegal copies of software are often an overlooked area of computer security. The Software Publishers Association (SPA) indicated in February, 1990, that during the preceding twenty months, they had begun lawsuits against thirty offenders of the federal copyright law. The U.S. International Trade Commission has estimated that copyright infringements total $4.1 billion annually. While the goal of SPA is for companies to establish active programs of employee compliance and education, there are still cases where legal action is necessary. Some companies having serious problems with the federal copyright law recently are:

Facts on File, Inc.—New York

Market Street Mortgage—Tampa, FL

Data Mark/Academy Insurance—Atlanta, GA

National Benefits Fund—New York

Hanoverfist Enterprises—Wayne, MI

Fox, S.A.—Barcelona, Spain

Cresvale Far East Ltd.—Hong Kong

Reproducing computer software without authorization violates the U.S. copyright law and is a federal offense. When you buy a single software product, the money you pay represents a licensing fee for the use of one copy. Civil damages for unauthorized software copying can be one hundred thousand dollars or more and criminal penalties include fines and imprisonment. When an unauthorized copy of a software product is used, it is not a pirated, bogus, or bootlegged copy—it is stolen property and should be treated as such. Companies can in no way, legally or ethically, condone, sanction, or approve the unauthorized copying of software.

SPA and the Business Software Alliance (BSA), an affiliate of SPA, are the principal trade groups of the PC software industry, representing over 625 members in North America and abroad. Through the SPA Copyright Protection Fund, the SPA and BSA have been instrumental in working to reduce software piracy. The SPA and BSA find out about companies violating the copyright laws in the following ways:

1. Honest employees call a vendor with a problem. The vendor requires the caller's name and company, and the serial number of the software product. If the vendor gets three or four callers with the same serial number, an investigation begins.

2. Contractors or visitors to the company observe evidence of illegal copies and then call the vendor.

3. Disgruntled employees call the vendor to report violations.

4. Honest employees concerned about illegal software in their company contact the antipiracy hotline at 1-800-388-PIR8.

Viruses are another security problem that have become common in businesses and require company procedures. Employees using home computers and modems or company equipment contact bulletin boards and get copies of public domain software. The software is generally a useful program that is free to anyone who wants a copy. Some of the programs, however, contain a virus that infects other programs by making copies of the virus and inserting them into noninfected programs. This insertion process takes only a fraction of a second, and is normally an undetected delay. The infected program will subsequently execute the virus code during its normal processing. The virus may case damage to programs and data or it may be relatively harmless.

PROTECT EMPLOYEES FROM UNNECESSARY TEMPTATION

A financial analyst working for the Washington, D.C., city government disagreed with his supervisors about where to invest the district's millions of dollars in cash flow. The analyst had unrestricted access to the financial database and password file. Tired of his superiors' interference in the investment process, the analyst changed the computer's access code and then "forgot" the new code; by doing so, he successfully blocked the district's ability to access its own funds. Additionally, the analyst started a guess-the-password contest in the local press and gave out daily clues. After a week, district officials were able to break into their own computers with the help of some high school students who had guessed the new access code. The analyst was fired, and instead of facing charges, he agreed to testify before the federal Grand Jury against his former supervisor on how city accounting subcontracts were awarded.

In other examples, a bored data entry operator who worked for an Oakland, California, department store changed delivery addresses so that thousands of dollars in merchandise was delivered to improper addresses, and three clerk-typists for a large midwestern manufacturer were arrested for running an "office" football pool on the company's word processors. The pool, which took in five thousand dollars a week, was part of a larger gambling operation that was being run on the mainframe computers and was clearing twenty-five thousand dollars a week.

INADEQUATE PROTECTION AGAINST DISGRUNTLED EMPLOYEES

In 1988, Donald Burlson received the first conviction for creating a computer virus; Burlson's virus deleted 168,000 sales commission records. In a similar case in a northern California software company, a customer support representative was fired but continued to access the company's computer system via modem. An investigation discovered that the former employee had copied company proprietary software worth millions of dollars.

Although loyalty to one's company is traditional, it has eroded over the years and the erosion is not just confined to regular employees. Two recent surveys show that job dissatisfaction now extends to managers and supervisors. One survey of middle managers found that over one-third of the respondents would be happier working for some other company, even if the new company didn't offer a salary increase. As a result, changing attitudes have to be addressed in the organization's security procedures, and changing attitudes make an overall security plan all that more imperative.

The typical computer violation is committed by an employee who is a legitimate and non-technical user. Often, the company's own policies lead employees to violate the rules. Some companies publish policies that restrict personal computers to company use. The problem with this policy is that when employees get accustomed to using the PC, they typically want to use it to write personal letters or college papers on their own time. If your company has such a policy, the following will result:

1. The employees will not use the PC to its fullest capacity. They will use it only for the specific, assigned task. Since these employees will not experiment to expand their knowledge of the new software packages, they will have a limited grasp of the technology.

2. Because they are so accustomed to using the PC, employees may just appropriate system time, and once they have violated company policy and seen how easy it is, they will be tempted to continue.

While most employees violate security rules for harmless reasons, others create serious financial damage. According to the FBI, the top reasons for committing computer crimes are:

- Personal or financial gain
- Entertainment
- Revenge
- Personal favor
- Challenge of beating the system
- Vandalism
- Accident

According to the FBI, a crime committed with a handgun nets the criminal an average of $19,000, while a computer-aided fraud nets the embezzler an average of $450,000. While the criminal with the handgun faces a 90 percent (or greater) chance of being caught and convicted, the computer-aided criminal's chances of being caught, tried, convicted, and serving any time in jail are less than 1 percent. More often than not, the successful computer criminal starts a computer security consulting business. One of the biggest reasons for the high return on computer-aided theft is that many companies try to handle the problem internally, and are unwilling to prosecute.

KEYS, COMBINATIONS, AND PASSWORDS CHANGED INFREQUENTLY

In October 1978, a contractor for the Security Pacific Bank learned how to use FED-WIRE, the method used to transfer funds from one bank to another. He opened a Swiss bank account and then waited outside the office of the vice president in charge of transferring funds until he left his office. The contractor then went inside, called the transfer operator and pretended to be the vice president. When the contractor requested that $10.2 million be transferred to the Swiss bank, the operator asked for the proper password for that transaction. The contract employee gave what he thought was the current password. The operator responded with, "No sir, that's the old password—here is the new password."

With the transfer complete, the contractor went to Europe, withdrew the money, and bought Russian diamonds, at which point the KGB contacted the CIA and told them of the purchase. With diamonds in hand, the contractor headed to his mother's house in Buffalo, New York, where the FBI met him. He was sentenced to eight years in prison, served three years, and was released on parole in 1982. He currently owns his own computer systems business in the Washington, D.C., area.

In many companies, audits and security reviews of offices turn up an alarming (and common) problem: Employees are leaving their desks unlocked. The problem is compounded because desk, office, file, and PC keys are usually kept in the unlocked desk drawers. As a result, the breakdown in security procedures allows for a new breed of thief: the key collector. This person collects any keys lying around and uses them in desks and offices that are locked. Many employees feel that they have nothing of value in their desks and therefore see no need to lock them; they fail to understand that they may assist in the violation of their fellow employee's property, not to mention the loss of corporate information.

POOR PROCEDURES FOR RECEIVING AND STORING EQUIPMENT

With the prolific growth of personal computers, there has been a parallel increase in the theft of desktop computer hardware and software. A 1985 survey of 184 of the Forbes 500 companies identified hardware theft as a three-million dollar-a-year loss. The survey indicated that 93 percent of the suspects in these thefts were employees or assumed to be employees.

To give an example, a family in San Francisco was arrested and charged in connection with the theft of 175 computers valued at nearly $1 million from Allstate Insurance company offices throughout California. The computers, which cost Allstate $6,000 each, were being fenced for $1,300.

Personal computers and related products such as modems have replaced electric typewriters as favored items to steal from schools, small businesses, and large companies. Current estimates of financial losses suffered by business are generally considered to be between three and five billion dollars annually, according to a 1990 U.S. International Trade Commission report, with less than 5 percent of all corporate computer crime reported.

EXPOSURE OF SENSITIVE INFORMATION IN THE TRASH

A new group of adventure seekers, the "dumpster divers," have applied themselves to gaining information about your organization or system by sifting through the trash. Remember, a printout can mean money and possible access to your company's systems and confidential information. Company phone books give potential interlopers employee names, departments, and phone numbers. As a result, you should treat company telephone books as classified information and dispose of them properly.

An example of how trash can be valuable occurred when two 16-year-old students in the Detroit area were arrested for credit card fraud. They went into the dumpsters around local hotels, restaurants, and department stores and then used the credit card numbers they found on the discarded carbon paper to run up ten thousand dollars each in purchases (mostly of computer equipment).

In a similar example, a California utility company threw out its old billing information and system access books until a high school student got into the dumpsters and figured out how to bill things to the utility. He rented a warehouse at the utility's expense and proceeded to fill it. Before he tired of the game and turned himself in, he charged two hundred thousand dollars worth of supplies.

Another example is that of a Los Angeles electronics wholesaler who gained access to the computerized inventory system of California's largest phone company using information he had gained from the company's trash. He diverted more than one million dollars in supplies to his business and, in some cases, sold the stolen equipment back to the phone company. He was eventually turned in by a disgruntled employee.

INADEQUATE OR NONEXISTENT SECURITY POLICIES

Before a program of employee awareness can begin, the company must first develop a formal position on information security. Published procedures are the first step in an overall information security awareness program. Without a published set of policies and procedures, a company cannot expect its employees to adhere to any standards.

ERRORS AND OMISSIONS

To be certain, there are outsiders, such as hackers, news media personnel, competitors, or curiosity seekers, who may try to access your company data, but the greatest threat to security comes from improperly trained or careless employees. Far more computer dollars are wasted than stolen.

To give an example, a design services supervisor submitted a series of three maintenance jobs (backup, delete, and restore). He was attempting to create additional space on the disk packs for the development of more design records. The jobs ran out of sequence (delete, backup and restore), and ten thousand design records were deleted from the graphics system. The supervisor then went to the data processing department for the system backup tapes to restore the deleted design records.

The problem did not end there. In reviewing the procedures, it was determined that the design group had changed some 2 disk packs from test to production status without notifying the data processing department. Since test volumes were only backed up on a quarterly basis, data processing could only restore 7,600 design records. The remaining 2,400 had to be restored using tapes that were almost ninety days old. It cost that company over one million dollars in overtime and lost production to restore the 2,400 back-level design records to current status.

And example of costly employee error occurred in a New York bank one Friday evening when a computer operator entered an incorrect code into the computer and billions of dollars that were supposed to be forwarded electronically to the Federal Reserve were left in the bank. Over the weekend this error cost the bank between ten and fifteen million dollars in lost interest.

RESOLVING THE DILEMMA

Achieve corporate commitment

As stated previously, management has the ultimate responsibility for protecting information from unauthorized access, and managers can be held personally liable for failing to protect information assets. When presented with this information, most managers quickly add their support to an information security program, and, recognition of their legal liability, combined with recognition of the benefits of the program, will win managers over.

Establish corporate policies

You should always publish the procedures and, wherever possible, keep them from being classified confidential. Your goal is to make sure the policies and procedures are read, so you should encourage people to make copies and attempt to distribute them widely. If possible, have the procedure manual placed online using any of the available text processing products (such as TextDBMS). Users will gain from having ready access to the procedures online, and you will be certain that they are accessing the most current generation.

Implement a security awareness program

Before employees can adopt company policies, they must first be made aware of them. It is not sufficient to just write the policies and procedures, the message must be communicated to the employees, and employees must incorporate it into their everyday work. Computer security cannot be a one-shot deal. To use a computer virus analogy, there is no vaccine for information security. An information security program is more akin to allergy desensitization: Employees must be exposed to information security on a regular and continuous basis.

Keep the message in front of the employees

Use whatever means available to keep the information security message alive. Some proven methods include posters, booklets, brochures, coasters, memos, reminder notices, and movies.

Monitor and audit compliance and results

Adopt an internal control review program to monitor the activities of each department. Although the audit staff will continue to perform regular reviews, the individual departments should be required to review information security standards on a more frequent basis. A questionnaire that addresses the key issues should be developed, and the questionnaire should be distributed and completed at least annually. If any of the questions receives a negative answer, then the department manager must submit a solution to correct the deficiency (similar to the response to an audit comment). The questionnaires should also be reviewed by the audit staff when the formal audit of the department takes place.

Make security compliance an appraisal item

At their annual appraisal, all employees should be evaluated on their compliance with all company policies and procedures, including those on information security. When employees are hired, most companies require new employees to sign a condition of employment agreement. Generally, the conditions include an agreement to abide by the corporation's policies and procedures. Since information security is a policy of the company, all employees should expect to be reviewed on how well they fulfill these requirements.

TYPES OF AGREEMENTS

All employees should be required to confirm that they have attended training sessions on protection of the organization's assets, and the confirmation can take many forms. Some examples of employee confirmation are provided in the following pages. The first is a general usage and responsibility statement.

The second is an employment agreement for employees accessing classified information.

USAGE AND RESPONSIBILITY STATEMENT

I understand that unauthorized use of or contribution to unauthorized use of computer facilities or company data constitutes a violation of established company policies. I recognize that I am responsible for maintaining the confidentiality of company information that I have access to while employed by the company and that failure to comply with these responsibilities is considered a violation of the Employee Code of Conduct.

GENERAL EMPLOYMENT AGREEMENT

I understand that my employment by the company creates a relationship of confidence and trust with respect to any information of a classified nature that may be disclosed to me by the company and that relates to the business of the company or to the business of any parent, subsidiary, affiliate, customer, or supplier of company-classified information. Such classified information includes, but is not limited to, inventions, marketing plans, product plans, business strategies, financial information, forecasts, personnel information, and customer lists.

At all times, both during my employment and after its termination, I will keep all such classified information in confidence and trust, and I will not use or disclose any such classified information without the written consent of the company, except as may be necessary to perform my duties as an employee of the company. Upon termination of my employment with the company, I will promptly deliver to the company all documents and materials of any nature pertaining to my work with the company and I will not take with me any documents or materials or copies thereof containing any classified information.

I represent that my performances of all the terms of this agreement and my duties as an employee of the company will not breach any classified information agreement with any former employer or other party. I represent that I will not bring with me to the company or use in the performance of my duties for the company any documents or materials of a former employer that are not generally available to the public.

I hereby authorize the company to notify others, including without limitation customers of the company and future employers, of the terms of this agreement and my responsibilities hereunder.

I understand that in the event of a breach or threatened breach of this agreement by me the company may suffer irreparable harm and will therefore be entitled to injunctive relief to enforce this agreement.

I understand that this agreement does not constitute a contract of employment or obligate the company to employ me for any stated period of time. This agreement shall be effective upon the affixing of my signature, namely:

Employee: _____

Dated: _____, 19 _____

EMPLOYMENT AGREEMENT FOR SECRET INFORMATION

Employee further agrees that, for a period of one year following termination of the employee's employment under this agreement, he/she shall not, directly or indirectly, engage or participate in or become employed by or render advisory or other services of the type encompassed by his/her former activities on behalf of employer; or acquire a direct or indirect interest (financial or otherwise) of greater than two-and-one-half percent (2-1/2%), in any firm, corporation, or business enterprise directly or indirectly competitive with employer.

Employee further agrees that, for one (1) year period following the termination of this agreement, employee shall not, without prior written approval of the company management committee, directly or indirectly, solicit or hire or cause to be hired any employee of or consultant to the employer, to engage or participate in or become employed by or render advisory services to any firm, corporation, or business enterprise in competition with the employer.

These agreements must be established and administered by the personnel department and should be part of orientation for new employees. The agreements should also be reviewed and reaffirmed by the employee on at least an annual basis.

BENEFITS OF SECURITY AWARENESS

Security awareness provides business controls, enhances efficiency and productivity, reduces theft, and improves safety and employee morale. Security awareness is a positive program that encourages employees to do the job right the first time—not just when it has to be done over. With a good security program, employees will be confident that the information used to make business decisions or to complete a project will be available and will have a high level of integrity. Additionally, the hardware that is needed to complete a job will also be reliable.

Responsibility for security

The employees are an essential element in protecting the organization's assets. The continued growth of the corporation depends on the reliability and trustworthiness of the employees. Generally, employees observe what management says and does. If management circumvents security, the employees will follow suit. To have an effective security program, all employees, including management, must approve of the security goals and philosophy.

BENEFITS VERSUS COST

To most employees, security is just a burden—pure overhead. Often, the value of security is not in what it brings to an organization (because it brings extra regulations), but how the organization would function without it.

The dollar cost of implementing a security program is similarly difficult to compute. Take disaster recovery planning or access control packages; both are major expenses, requiring an initial investment of time and money and extended, long-term administration and maintenance. Too often the only cost factors seen by management are the accounts payable. When security is functioning well, there does not appear to be any bang for the buck. However, if your organization ever has to put its disaster recovery plan into practice, has classified information leaked to the competition, or gets invaded by hackers, then the value of a security program will become very apparent.

When developing policies and procedures or implementing a security program, always maintain a business perspective. If there is a security breach, the key questions are: does the organization *want* to recover? and, *can* it recover? The goal of your organization's security program includes:

1. Prevention—the security program must be able to prevent as many security failure problems as possible.

2. Detection—because no security program can prevent 100 percent of the security problems, sufficient controls and reporting mechanisms must be in place to detect any problems or possible intrusions as quickly as possible. Detection includes implementing audit trails to trace the event to the source.

3. Containment—once a security breach has been detected (or, even more importantly, before a security breach occurs), controls must be in place to contain the problem. With currently available security packages, containment can be accomplished relatively automatically.

4. Recovery—the final step is to restore the organization's computer facility back to the pre-incident level and to ensure the integrity of the organization's information.

IMPLEMENTING POLICIES AND PROCEDURES

The subject of how to create policies and procedures and how to sell them to management has been discussed at great length. The next section describes ways to disseminate the new policies to the general employee population. Your objective is to ensure that the formal roll-out of your documents will mesh with the procedures already in effect and will suit the style of your organization. Like your mission statement, you should write a cover letter and allow for a grace period and training, as necessary.

The cover letter should briefly describe the need for the new program. By discussing the need for the program, employees will see that the policies and procedures were developed for a specific purpose and that the overall goal is to improve the working conditions and business climate of your organization. The cover letter should be signed by the highest possible level of senior management.

For example, an increased level of security was recently introduced at a campus facility and the transition went very smoothly. Representatives from each department were invited to attend a meeting on the topic of increased security. This group of employees usually met once a month to review on-going programs, and they were used to being called in to receive news of changes in operating procedures. At the meeting, the new program was introduced and the reasons for the implementation were given. Each of the attendees was given a formal notice signed by the head of personnel. This letter reiterated what was discussed at the meeting, and copies were provided so that each employee in the respective departments could receive a copy. A general question-and-answer period followed the distribution of the letter. After all questions were answered, the attendees headed back to their departments to inform the other employees of the new program. You should consider this kind of introduction when implementing your security program. Again, factor in the background of your organization. A memo from the boss may be sufficient. For the best results, however, meetings are generally much more effective in communicating objectives, particularly to a large corporation.

The second issue to consider is whether a grace period is appropriate. Once the new policies and procedures are published, you will need to decide when they will go into effect. One proven method is to require the departments, units, and divisions to develop a compliance plan and submit the plan to a central office by a specified date. This kind of action causes the recipients to read the new manual, determine areas where they are deficient, establish a plan of compliance for each situation, and identify when they will be in compliance. By requiring a compliance plan, the organization ensures that the new policies and procedures have been read, and the departments have responded to them. By setting a deadline for completing the compliance plan (usually two to three months after publication), the unit is given a grace period from audit comments on the new material.

The third element is to determine if employee awareness and compliance training are necessary. Awareness programs are discussed in depth in Chapter 7, but for now, you should probably assume the need for an on-going awareness program and plan accordingly.

When the manual is published, make certain that it is easily accessible. Currently, most procedures are published in printed form, but many progressive companies are moving to on-line documentation. Whatever the format, remember to advertise the manual's existence. Company newsletters, staff meetings, lunchtime newsletters, and new employee orientation are all good mediums for communication. Your goal is to ensure that no employee can plead ignorance as an excuse for noncompliance.

AVOIDING PITFALLS

As a computer or information security officer, you already know that your job is characterized by sniping and attacks from all sides. No one in your organization wants to see you enter a room, section, department, or building. All too often, you, the security officer, are left alone to figure things out for yourself. In many organizations, you are the first employee to attempt an effective computer security program. Your experience, therefore, seems unique, but your job is prone to the same pitfalls faced by other computer security personnel. Some of those pitfalls are described in the following list:

1. *"The job of information security officer is a fast track to the top."*

If this is what you believe—don't take the job!

If you like a challenge, then the job of developing a computer or information security program is for you. You must be forewarned, however, that none of your fellow workers wants you to succeed, because, for you to do your job, others will have to change the way they do their job.

To do your job correctly and efficiently, you will have to become a great compromiser. Your progress, furthermore, will often come in very small increments. As soon as you succeed in one area, such as publishing security procedures, users will be on the lookout for loopholes to get around those procedures.

2. *"If we can find the right hardware or software, then all of our security problems will be solved."*

Management often believes that the wizards in data processing can reach into their bag of tricks and produce instant security. Security is not a technical problem, but a management or people issue. The tools exist to create a secure environment both in hardware and software. The problem is that when people are added to the equation, security often gets compromised. Management must support the concept that information security must be incorporated into the daily activities of all employees.

3. *"Let's form a committee to study the issue. Since the concepts of computer and information security are so vast, we'll need to establish subcommittees that will report back to the central committee on a regular basis."*

Overplanning is a delaying tactic. In government, when an administration is forced to address an issue, it establishes a blue-ribbon panel to look into the problem. When you know what the problem is and you have the ability to correct the problem, and the benefits outweigh the costs, then fix it; don't just plan to fix it.

This doesn't mean you shouldn't plan. It does mean that management expects results, not just a plan.

4. *"We've got a plan, and, by George, we're going to stick to it."*

You may be familiar with this attitude if you've ever tried to assemble some toy for your children on Christmas Eve. As the night wears on, and you grow more tired and the parts don't fit—you just reach back and—force it!

Even if you have the greatest plan in the world, circumstances may change so much that you have to modify the plan. For example, if your plan calls for installing an access control package within a specified time frame, don't set it up so that the package for the entire user community is in full abort mode after a long weekend. Start small, perhaps within your own department. Iron out the bugs, compare the problems encountered with the problems you had anticipated, adjust your plan, and only then move on to another department.

5. *"The three biggest lies heard in the data processing departments are:*

• *This will only take a minute.*

• *This will be transparent to the user community.*

• *Hi, I'm from the audit staff, and I'm here to help you."*

The auditor is not your enemy: quite the contrary. Both you and the auditor have the same objectives, to ensure that internal controls are sufficient to protect the company from loss or risk.

To make your life easier, you should establish a liaison with your audit staff. The audit staff can give you a feel for what is going on in the company—what programs are working and where security measures are inadequate. And remember, it will be your job to develop the procedures that will be used by the audit staff when they begin to monitor compliance.

6. *"This plan speaks for itself. There's no need for me to go out and be a security salesperson."*

Wrong! The easiest job is to create the policies and procedures necessary for the overall security plan. The toughest job is to make management and the employees aware of the security plan and their role in supporting it.

Managers generally do not have the time to read all the procedures. If you want your program to be supported by management, you're going to have to get out of your office and present your case at every staff or department meeting you can.

Good managers hire good people and then they leave them alone to do their job. Managers will expect the ISSO to take the initiative. Present your case, point to your accomplishments, sell your program, and management will support your actions.

7. *"Publish or perish. If you can't win them over on substance, then bury them with volume."*

There is a law of diminishing returns when it comes to the generation of paper: Less is more. Dumping a three-inch violation report on the purchasing manager's desk and expecting her to search through the printout to find the violation will prove counterproductive.

The job of the ISSO is to develop procedures to meet company needs and then motivate the employees to follow the procedures. Motivation can be developed by a variety of awareness programs and a company information security certification program for each department.

8. *"The ISSO! Who came to the company with powers far beyond those of mere mortal employees! Able to install a new security package in a moment's notice! Hurtle security problems in a single bound and research the latest computer and information security laws after work hours!"*

No, you aren't superman or ever Rick Tracker, DPI. You can't do it all by yourself. Seek out the experts within your company and involve the end-users—after all, what you create is going to affect the way they perform their jobs, and you need their active participation.

You are not the sole guardian of company security. Each employee is responsible for his or her own actions and that includes protecting company information and computer assets.

9. *"Computer and information security are our organization's most important product."*

While security may be *your* number one priority, the rest of the employees are engaged in carrying on the business of the company. You must strive to provide the maximum security with the least impact and lowest cost.

Although the most secure computer is one that is down, a computer that is down is not in the best interest of the organization. The computer will have to be operational and the employees have to have access to the system and data if the business goals of the organization are to be fulfilled.

When there is a choice between generating revenue or fixing a security breach, you'll probably have to come up with an alternative to security.

10. *"ACLS are required for files, IPC objects, and UNIX system domain sockets. Access control for sockets that use name spaces other than those to the UNIX system (UDP, TCP) must be addressed in the specifications and evaluation of the system involved."*

Speak and write in plain English. Don't write in computerese or use buzz words or bureaucratic doublespeak. When in doubt about your audience, think of how you would explain technical jargon to folks like your parents or others who have limited exposure to computer terms.

Also remember that a policy not explained is a policy not followed. Never assume that the words speak for themselves. The employees and managers will benefit from discussion and interpretation of new policies. Be certain to anticipate these needs and provide additional explanatory information and documents.

CHAPTER REVIEW

1. Security is not a technical problem; it is a management and people problem.
2. There are three separate groups to which you will have to sell your policies and procedures:
   • Management
   • Data processing
   • Users
3. Before approaching senior management, determine the current level of security at your organization by:
   • Using audit reports
   • Conducting a walk-about
   • Checking theft reports
4. Keep management informed of your progress.
5. Gaining support from management will generally be manifested differently:
   • Senior management = funding
   • Middle management = resource help
   • First-level management = compliance
6. Employees must be made aware of their responsibilities for compliance.
7. The benefits of a security program must be presented in business objectives:
   • Provide consistent controls.
   • Enhance efficiency and productivity.
   • Reduce theft.
   • Improve safety and employee morale.
8. The goals of a good security program are:
   • Prevention
   • Detection
   • Containment
   • Recovery

9. Publish the security manual in an easily accessible medium.

10. Remember to avoid the pitfalls.

Exercise # 1

Using the general reasons given for security failures, identify any additional reasons why security might fail at your location.

- Develop an illustrative example.
- Create a response to correct the problem.

Exercise # 2

Using the business objectives discussed as positive selling points for your security program, develop additional points. Consider the following:

- Reduction in overall employee workforce.
- Compliance with regulatory agencies.

CHAPTER 7

Employee Awareness Program

As stated previously, publishing a set of policies and procedures is no assurance that anyone will ever read them. Creating an employee awareness program is really necessary to bring the security message home to the employees. Before an organization can accept a security program, however, the employees must first understand why the program is necessary and what they will gain from its implementation.

Each group of employees within your organization will have a different focus on the need for a security program. Application programmers understand the need for structure and controls to complete their assignments in a timely manner. The operations staff understands the need for separation of duties. Senior management understands the need for security but often is not fully aware of all of the exposures in the organization. The end-users only understand the need to get their jobs completed and look upon security as an impediment to that goal. While different areas of the organization understand different aspects of security, few put all the elements together.

Although a program can never be all things to all people, the employees must understand the need and benefit of the program. The goal of awareness, therefore, is to convince the employees of the benefits that the program has to offer.

PROBLEMS IN SELLING EMPLOYEE AWARENESS PROGRAMS

When you discuss the need for access control packages to protect the organization's data, many top management personnel will point to the fact that the computer room is protected by a very expensive card access system. The concept of computer access from remote locations, both within and outside the building, can often be lost on management—who may not be as computer literate as you would expect. Senior managers of most organizations are the last bastion of people who grew up before computers. In most executive offices today, the use of terminals or personal computers is still restricted to the secretary or the most junior member of the management team. It will take until the turn of the century before computers will be fully integrated into executive offices.

Because the scope of a security program can be enormous, many security officers get overwhelmed by the job and attempt to implement a program that is too comprehensive before the organization is ready for it. The key to creating an effective awareness program is to start small. Find the most pressing need and then address it.

Another problem with creating an employee awareness program is dealing with a limited attention span. The average employee has many things on his or her mind during a meeting. You will have to structure your presentation to grab the employees' attention as quickly as possible and then provide them with the information that they can use to complete their job assignments. Never create a presentation that is too technical. Do you really need to tell management that ACF2 has been moved into abort mode? Would managers understand encryption better if you gave them a detailed mathematical explanation of the Data Encryption Standard (DES) algorithm? Using too much technical detail will only turn off the very people that you will count on to accept and support your security program.

Another problem in selling security is simply to get the employees to read the policies. One proven method is to develop a booklet or small pamphlet. A booklet or pamphlet can usually be developed in-house and can provide the perfect vehicle for communicating the important points of the security program to the employees. This booklet should be distributed widely throughout your organization. There are two schools of thought on how this information should be disseminated. Some feel that the booklet or pamphlet should be mailed to the employees or just dropped off on their desk at night. Others feel that each employee should be personally given a copy by some member of management or the security staff. Either of these methods can be effective, and it's important not to focus all your effort on how to distribute the material. Remember, this initial consciousness-raising effort is only the first step of the ongoing awareness program.

Once the employees have been made aware of the existence of the security policies and procedures, they must be made aware of how the new controls will affect their jobs and what their role is in the overall program.

THE SCOPE OF AN AWARENESS PROGRAM

An awareness program is actually a sales pitch, similar to selling the security program to upper management, except that the focus for the awareness program is the general employee population. When directing the sales pitch to management, the emphasis was on establishing controls to support the business process and management's fiduciary responsibilities. When addressing employees, the awareness program must emphasize the benefits to them personally and to the organization as a whole.

Since security is part of their job description (or a key element thereof), most employees have already had some exposure to what is expected of them. Typically, however, most employees focus on getting their job done. The goal of an awareness program is to remind employees of the control elements in their job descriptions and to assist them in integrating the elements into their day-to-day activities. The awareness program you develop should engage your audience and trigger an automatic recall of the security message whenever employees face the issue in their workplace.

TAILORING YOUR PRESENTATION

Unlike the political speeches of some candidates, you will have to vary your material and adapt it to the different groups in your organization. As you begin to present the awareness program, you will find out that the organization is made up of many different areas and that one stock presentation will not play well in all of them. To be effective, your program should be designed to meet the needs of the various user groups.

To ensure that the message gets to the employees in a manner that they find acceptable, you may want to consider a train-the-trainers approach, where you train a selected group of employees who then train the end-users.

By training employees in each of the business areas, a number of positive elements result. One advantage of the train-the-trainer approach is that you will not have to train all of the employees in your organization. If you have a small company, this may not be an issue, but with a larger company it may be impossible to train everyone yourself. As an example, when I began the awareness training program for my company, there were 813,000 employees in 291 locations in 33 states and 23 countries. From February to September of the kickoff year, I totaled 33,000 domestic and 18,000 overseas air miles, conducting 55 meetings in 14 states and 5 foreign countries. A total of 477 employees were trained to take the security message back to their place of employment.

During these training sessions, the representatives from each of the areas were given a basic awareness presentation kit. The kit included a set of procedures, a set of awareness posters, and a video to reinforce the message. To assist in the process, the video and some of the handout materials had been translated into Spanish, French, and German. I felt very well prepared when I was setting up the South American training session in Brazil because I could tell the host organization that I had copies of the materials available in Spanish.

Unfortunately it was then that I learned that Portuguese is spoken in Brazil. This story is a good reminder that a lot of work will have to be done to prepare the awareness materials for the different locations and groups within your organization.

The proximity of a location to the organization headquarters is also an element that must be incorporated into the awareness program. In the security business, this problem is called the flagpole versus the boondocks syndrome. Geography plays a big factor in how well controls are followed. If you have a remote location that is visited by security and audit personnel only on an occasional basis, then the compliance will reflect the distance. The further the location from the main site, the less likely that controls will be followed. This kind of problem can be corrected by a concerted effort and by regular visits to the outlying sites. Since they are part of the organization, they must be treated just like any department or section at organization headquarters.

Finally, you should be aware that in every organization there are at least a few problem groups: senior managers, engineers, and personnel staff. The presentation for each group should use the vocabulary and terms that are common to the particular group. As discussed before, the senior managers are generally interested in the bottom line. The presentation made for this group should be short, to the point, and show how the program will improve the business process.

The personnel staff will also show some resistance to the awareness training program. Personnel employees are always running off to other meetings, and it is very difficult to get them to make the commitment to attend training meetings and become actively involved in the control process. Many personnel department employees look upon computer and information security as a data processing problem, and they don't see that it benefits them. Your job will be to structure the presentation so they recognize that the security program will provide them with the controls needed to protect the sensitive information maintained by the personnel offices.

Engineers and system engineers are another problem group. The chief obstacle with these employees is their work methods. They are so busy trying to solve the problems of the organization or the computer system that they are always on the lookout for shortcuts. This element of their job leads them to discover ways to increase the productivity of the organization and the business process as a whole. However, their free spirit approach leads them to disdain controls of any type.

One method to get engineers and systems programmers to listen to the security message is to play on their territorial instincts. They often become very protective of the information and data that they have created. By making them aware that controls will keep other employees away from their work, the engineers and system programmers will begin to adopt security controls on their own. In fact, many times I have gone back to their areas and found that the level of control is much higher than it is elsewhere in the organization.

AWARENESS TRAINING AS AN ONGOING PROCESS

All too often, a new program is introduced to employees with a big initial splash and then the program is never heard of again. When this happens, the employees look at a new program as a temporary inconvenience. With that in mind, you must make sure that any new training program is planned along with a regular follow-up procedure.

While the awareness program requires a continuous, steady flow of information, you must make sure that the employees are not overwhelmed by the new material. Keep your message in front of the employees, but practice good timing. Remember that you can also use the press as a means of getting your message out to the employees. If a news article describes a security problem that is addressed by your program, take the time to notify management of the news article and explain how the security program will protect your organization from the kind of problem described in the article. Always try to stress the positive side of security and let everyone know that the security program is really about protecting the company assets.

The most effective program has three phases. The first phase is a formal introduction in which a key executive presents a three- to five-minute outline of the objectives of the program. The message by the executive should be taped for use at all the initial presentations. The second phase should be the follow-up procedure. This would include desk drops, such as pamphlets, posters, and newsletter articles on the subject. A desk drop is normally conducted at night and in some organizations is done by the security staff as part of its nightly rounds. The final phase is the employee review in which adherence to security controls are factored into the employee's appraisal. For many employees, this is when the reality of the security program hits home. When their compliance with security requirements directly affects their promotion and pay increases, most employees will get behind the program.

MAKE MATERIALS CONSISTENT

To prevent mixed messages for the employees, each handout should be consistent with the general policies of the organization. Whenever a new reminder aid is developed, try to use the policy wording to reinforce the direction you have been taking. The goal here is to send the employees the message that the program is still in effect and that the level of expectation has not changed. Once you establish the direction of the program, the effectiveness of the program will hinge on how consistently is is presented. If the employees perceive that there is a change in direction, there will be inevitable disruptions in the work environment.

LISTING THE PRIORITIES FOR PROTECTION

The employees need to understand, in concrete terms, what is to be protected. To gain the employees' understanding, you should list the assets that must be protected in the order of their priority. The number one asset to be protected is, of course, the employees. To gain their support, the needs of the employees must always be addressed first.

Once the employees have been adequately protected, the next most important element is data and information. While some may view cash reserves as more valuable than data or information, it's important to recognize that if information is lost, your organization may not be able to generate any cash at all. Think, for example, about the loss of all the accounts receivable information. If your organization lost the ability to collect the money owed to it, the long-term health of the organization would be jeopardized. Therefore, data and information normally rank ahead of the cash reserves as organizational priorities.

Business integrity should be next on the list. The way your customers perceive the company is an essential element in its long-term survival. A loss in customer confidence can spell disaster for a business, whether it's in manufacturing, services, or the health industry. What would be the effect on a business if the customer was unsure of the product being provided or on the organization's ability to service the products provided? The consumer would probably turn to your competitor. With proper controls on information access, the integrity of the products and services provided by your organization will meet the demands of the public.

The next list item to be protected must include the physical assets: the buildings, the computer hardware and software, and the physical environment. Employees need to have a place to perform their job functions free from distractions and harm. The physical buildings need to be protected from not only natural disasters, such as fire, hurricane, flood, and earthquake, but also from terrorist activities. When controls over the physical plant are firmly in place, employees can concentrate on improving the business process and answering the customers' needs. Increasingly in the past decade, the business community has recognized the need for safeguarding physical security and developing awareness for this aspect of your program has actually become easier over time.

The last item on your list of protection priorities should be the ability to monitor and audit the effectiveness of the controls that currently are required in your organization. Employees must learn to view an audit as a positive business activity. True, the auditors sometimes point out areas of deficiency, but they also check to see that all assets of the organization are being given adequate attention and protection. The audit process must be viewed by all involved as a learning tool that will provide for better controls and an improved product.

The goal of a security program is protective, not punitive. An effective awareness program will make the employees understand that the continued growth of the organization depends on how well they do their jobs. Any presentation that you develop to sell the security program must stress the positive aspects. You will, of course, have to take a little time to identify the penalties for noncompliance, but the main goal of an effective program is to stress the benefits.

CONTROLS IN AN EFFECTIVE PROGRAM

Controls are necessary to ensure that all of the organization's assets are protected from loss, including the physical property such as personal computers, printers, plotters, and software. The most damaging to any organization, however, is the loss of its data and information. When a disk is stolen, the replacement cost may only be a dollar or so, but the data on it is usually far more valuable. Through the years, I have investigated a number of reports about missing disks, and, for the most part, the disks were not taken for their information value. No matter how many hours were spent developing the information or how confidential the information was, the disks were usually just reformatted for home or school use.

Despite the preceding information, you should not eliminate the concern for disclosure of information from your control requirements. Information disclosure is still a security issue, but when you restrict data access to those with a clear need to know, unauthorized disclosure of information becomes less of an issue. When information is disclosed, it is almost always an employee who discloses it. Industrial espionage has occasionally been a factor in information disclosure, but restricting access on the need to know principle reduces the chance that your organization's information will be given to the competition.

By controlling access to information, the chances of unauthorized alteration of data are also reduced. The audit trail requirements also support the existing controls by monitoring classified data and critical systems to detect any alterations. In computer security, unauthorized modification of data is the biggest concern. The most frequent criticism is that too many employees have access to data and have access at a higher level that is necessary to do their jobs. By establishing access control requirements and ensuring that the data owners monitor the access elements, the risk of exposure and unauthorized modification of data is reduced. The positive side of these controls for the users is that the data they access will not be corrupted and will allow them to make informed business decisions.

An effective awareness and control program will also assist in protecting the company from loss of computer services. While unauthorized access to mainframe computer systems has traditionally been the focus of security programs, personal computer use is an area of increasing concern. The controls that exist in the mainframe world are often missing from the PC environment. An effective awareness program will present PC users with a positive set of controls to ensure data integrity. Additionally, by installing a physical device on all PCs, your organization may be able to stem the tide of theft of PCs and peripheral equipment.

UNDERSTAND PRESENTATION TIMING

Before you begin to schedule security awareness presentations, you must be sensitive to the timing for the various departments. The accounting department is not going to be interested in attending a security awareness program during the month-end accounting process. The payroll staff has weekly busy periods, and the stores and inventory sections are not likely to attend a session during their annual inventory process. So be sympathetic to each department's schedule and how it relates to your organization's yearly cycle.

The time of day for a meeting is also vitally important. It is strongly recommended that you *never* schedule a security meeting just after lunch. When I taught high school, the schedule was set up so that the students who were not on the college track would have wood shop, metal shop, auto shop, gym, lunch, and then take my class on American history. There is nothing like a seventeen-year-old who has run around all morning, just eaten lunch, and is now just looking for a warm dark place to take a nap. The eyes of the students would start to flutter after about ten minutes of class. For the rest of the hour, it was a race to see how much information I could give them before they nodded off to sleep. Your fellow employees have worked hard all morning, they just had lunch, and now they have only two concerns: keeping their lunch down and finding a spot to catch a few winks. Right after lunch is a bad time to schedule a meeting.

Additionally, if your organization has a third shift, offering the employees a hour of overtime to stay and listen to the awareness presentation is not inducement to stay awake. Try to go to the employees and work your presentations around their schedule. By taking into consideration the employees' needs and work schedule, you will present to them a positive message of understanding.

The best time to schedule security awareness meetings is on Tuesday, Wednesday, or Thursday mornings between 8:30 and 11:30. This will give you the freshest audience and will give the awareness program the best shot at success. Remember, try to be flexible as possible.

TARGET AUDIENCE

Because this program is based on the protection of all the organization's information (not just the data stored on computer systems), any employee with access to the organization's information should be included, whether the employee is salaried or hourly, full- or part-time. Contract employees may also be included in the awareness training, but check with the personnel staff for guidelines in this area. Over the past few years, there has been a growing concern about treating contract employees as regular employees, particularly when the organization has a long-term relationship with a contract employee. So, to be on the safe side, check with your personnel staff for direction before scheduling any contract employees in your security awareness sessions.

TYPICAL AWARENESS PROGRAM

The time of day, week, and month that the sessions should be scheduled have already been discussed. The normal session should not last any longer than one hour. For most efficient scheduling, try to have each session run about fifty minutes and use the remaining ten minutes of the hour to clear the room and bring in the new group.

The typical session should begin with an introduction of the topic. If possible, try to play a taped message from a high-level executive. The introduction should be followed by a video on the topic of information or computer security. A listing and brief annotation of appropriate videos are included in the Appendix.

After the video, you should discuss any methods that will be used to monitor compliance with the new program. Before you begin a program to check compliance, you should always notify the employees about the program. During this time, you should explain the reasons for checking compliance and explain the consequences of noncompliance. Any program that is discussed must be implemented. If a compliance monitoring program is announced at the training session and then nothing is done, the employees will get the impression that the security program has no clout and therefore can be discarded.

Finally, pass out a booklet that provides a summary of the items covered in the awareness program. This booklet should make reference to the new policies and procedures, describe where copies of those documents can be found, and mention whom they can contact for additional information and training.

Often, organizations take this opportunity to present the usage and responsibility statements or an attendance verification form indicating that the employee has attended the training program and has had the opportunity to ask questions. The verification form reinforces the concept of employee responsibility for protecting company information.

CHAPTER SUMMARY

1. Stress the positive aspects of how security affects job duties, company profitability, and employee rewards for such profitability.
2. Design your program around a strong theme and make it interesting.
3. Make it clear to the employees that the information and computer security program has the support of top management.
4. Provide your audience with the message and materials that they can relate to—something that they can take with them and refer to later on.
5. Use the wording of the security policy statement as the basis for all written materials.
6. Make your program multifaceted by using a variety of materials, techniques, and media sources. Varying the approach increases the chances that you will reach your intended audience.
7. Stagger the release of the materials. The awareness program cannot be just a one-time event; it must be part of the employees' work environment. Deliver posters to be put up in office areas regularly. Drop pamphlets, brochures, and booklets at the employees' desks at least four times a year. These materials will reinforce the message for employees.

Exercise #1

In order to develop an effective awareness program, you must determine the topics to cover in your initial awareness presentation. Make a list of those items that require immediate attention and assign a priority to each.

Exercise #2

Establish your target audience. Identify any specific job functions that may not benefit from the awareness program.

Exercise #3

Identify the departments or sections within your organization through which your training program should be coordinated. This may include such sections as education and training, communications, public relations, and labor relations.

Exercise #4

How would you handle the training requirements for contract employees? Should they be part of the regular employee training, or should they be separate?

CHAPTER 8

Monitoring Compliance

The importance of formalizing management controls cannot be overemphasized. Failure to provide written guidelines results in continuing employee misunderstanding and confusion. Additionally, a lack of formal controls results in operating inefficiencies and may make the organization subject to litigation. After you have developed an employee awareness program and begun your follow-up training, you must develop a program to monitor compliance with the published policies and procedures. Figure 19 shows the phases of control development.

The quality and effectiveness of an organization's internal controls are directly related to how much management stands behind the policies and procedures. An effective and aggressive management team creates an atmosphere that gives the message that controls are important here. If control is only given lip service or is contradicted by the actions of top management, then poor control will result.

A major East Coast telecommunications corporation had a policy that all employees wear identification badges whenever they were on the company grounds or in the building. After about a year of this policy, the president called the head of security into his office and asked in the company had a policy on employee identification. The security chief responded that the policy had been in effect for quite some time. The president then mentioned that he had been wandering around the building that day and had noticed that very few employees were wearing their badges and he asked the security chief for an explanation. The security chief responded that the employees weren't wearing their badges because the president didn't wear his badge. The security chief said, "the first group to stop wearing badges was your staff, and then the members of the other staffs stopped wearing their badges. Soon all of the managers and important employees stopped wearing badges. The only ones currently wearing them are those who work for security or those who are too new or too dumb to know what is going on around here." From that day on, the president wore his badge and he made sure that any employee or visitor he came upon had their badge displayed.

FIGURE 19 **PHASES OF CONTROL DEVELOPMENT**

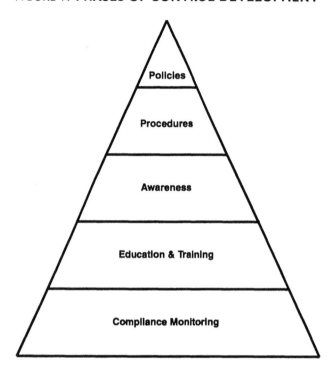

CONTROL ENVIRONMENT

The first step in monitoring compliance is to designate high-level responsibility for control. Your policy should therefore state that senior managers or appropriate heads of staff are responsible for establishing controls within their areas to protect the organizational assets from unnecessary risk.

Managers should regularly review the status of the controls and provide at least semi-annual updates to the management staff. The update should be a formal presentation and include specific examples of where controls are inadequate and recommendations on how to correct the deficiencies.

To assist the employees in understanding their roles, an employee code of conduct should be published. The code, usually in pamphlet or booklet form, should outline the employee's responsibilities for compliance to all laws, regulations, policies, and procedures. The code is also valuable as a public statement by the organization on its business principles. To be completely effective, the code of conduct should state the measures that may be taken if an employee violates the guidelines.

During the employee's annual appraisal, the employee should sign a conflict of interest statement. The conflict of interest statement serves to remind the employee that he or she should avoid an activity, investment, or interest that might reflect unfavorably upon his or her integrity or good name or that of the organization. Any employee agreements should also be reviewed during the annual review. For more information on employment agreements, see Chapter 3.

An effective system of authorization approval must also be established. A system of authorization approval will first of all establish the owner of an asset. Once the owner is established, the employees then know who they must reach for access authorization. Nothing slows down a job assignment or irritates employees faster than not knowing how to get access or whom they must ask for approval. By standardizing the approval process, job efficiency will increase and employee morale will improve. And with the levels of authorization clearly established, the vehicle for monitoring controls has been created.

An informal, internal review program is a good way to ensure that controls are being met. A series of questions to be answered by the various departments will provide the basis from which the organization can report the level of compliance.

INTERNAL REVIEW PROGRAM

The broadest goal of internal controls is to satisfy the organization's overall business objectives. One approach is to consider the major risks that an organization will face and then set some objectives for minimizing these risks. The review program should not try to complete a full audit of each area, but should simply identify weaknesses and allow the department conducting the review to implement corrective actions. By enlisting the active participation of the local manager, the review program will minimize occurrences, but will not completely eliminate all risks.

SAMPLE QUESTIONS

In order to get the local managers actively involved in the review process, your organization will have to come up with a series of questions that will allow the manager to evaluate the current level of control in his or her department. To provide the best possible results, the questions should be as straightforward as possible and should include a reference to the corresponding policy in the security procedures manual. The following examples will give you some guidelines for how questions may be grouped and what information may be included.

Information Security Standards

1. Have you read and do you understand the contents of the information and computer security manual?
2. Have you implemented the prescribed standards?
3. Have you noted and documented variations from the security policies?
4. Have you noted and documented any other security exposures not covered by the security manual?
5. If you have noted a variation from the security manual or some significant security vulnerability, what steps have you taken to resolve the situation?

Data Classification System (DCS)

1. Do you understand the DCS?
2. Has the information within your area of control been classified?
3. Has an information owner been identified and has this designation been forwarded to the information security administrator?
4. Are you taking the required steps to ensure that proper protection is given to all company information within your area?
5. Have all employees within your area been reminded of their responsibility to use information only as authorized by the owner?
6. Is the level of protection for classified information reviewed periodically to ensure its adequacy?

Employee Security Awareness

1. Do you periodically remind your employees of their information security responsibilities?
2. Has information about information security awareness been disseminated to everyone in your department?
3. Have you implemented a security awareness program within your area of supervision?
4. Do you periodically monitor your employees to ensure compliance with organization standards?

Other Security Measures

1. Have you identified all key assets (such as computer hardware, software, telecommunication hardware, and information) within your area of supervision?
2. Have you identified actions taken to protect key assets?
3. If any key assets are not adequately protected, has a decision to that effect been justified, documented, and formally approved?

SECURITY-AUDIT ALLIANCE

While an informal review by department managers is a positive and necessary part of the overall security program, the security and audit staffs must ultimately be responsible for checking compliance. The information security officers and the audit staff have always had a great deal in common; the two disciplines should be traditional allies. However, through the years, the two sides have functioned more as enemies than as allies. This relationship between the two is not fostered when the security officer and the auditor only get together when the auditor arrives for a formal visit. This lack of communication results in an adversarial relationship and often leads to downright hostility.

While it may seem like a conflict of interest, I recommend that you go to lunch with an auditor on a regular basis. You cannot develop a good working relationship if the only time you see each other is during the audit process. Take time out to get to know the audit staff. The auditors have great insight into what is going on in the organization and can give recommendations on what policies and procedures are working and what ones may need a little work. The next sections will provide some background on the roles of the auditor and the security officer.

THE ROLE OF THE AUDITOR

An audit staff mission statement or charter (see Chapter 2 for more details) might include the following:

The objective of the audit staff is to report on the adequacy of the controls and protection of the organization's physical and information assets at all locations. The audit staff will identify and recommend areas of the organization to be examined, develop audit plans based on risk analysis data, conduct examinations, and submit reports of their findings with any recommendations to senior management. The audit staff will review local- and organization-level compliance with procedures and policies of the organization and will examine the elements of the organization's internal control system. The audit staff will be responsible to senior management for the following:

- Providing an independent and objective appraisal of controls.
- Ascertaining the extent of compliance with established policies, plans, and procedures.
- Determining the extent to which organization assets are accounted for and safeguarded from loss of all kinds.
- Ascertaining the reliability of data used by management to make business decisions.

• Reviewing the quality of performance in carrying out assigned responsibilities.

• Recommending operating improvements.

• Providing the organization with advisory-type services as needed.

Using the mission statement you created as an exercise in Chapter 2, you can compare the goals of the audit staff with those of the security officer. The common thread in both groups is the concern for establishing a proper control environment throughout the organization. The following chart gives a quick comparison of the audit and security roles.

Because the security staff holds a position of considerable trust in the company, the auditors must be very rigorous in examining the data security department. Individuals in both the audit and security groups are held to a higher ethical standard than the average employee. The security personnel, in particular, are expected to adhere not to just the letter of the policy or procedure, but actually to the spirit or intent of the control item. Security and auditors should not use loopholes, but they are required to find them, report them, and fix them. Auditors can help ensure that the information and computer security personnel maintain a high standard of integrity.

RESPONSIBILITY AND AUTHORITY

The responsibilities of the organization's audit staff have usually been established by senior management over a period of time. The policies generally give

FIGURE 20 **COMPARISON OF ROLES**

AUDIT	**SECURITY**
APPRAISE CONTROLS	**IMPLEMENT CONTROLS**
● AUDIT	● SECURITY
● EXAMINE POLICY COMPLIANCE	● CREATE POLICIES
● ASCERTAIN CONTROL ADEQUACY	● ADVISE ON APPROPRIATE CONTROLS
● REVIEW DATA PROCESSING DEVELOPMENTS	● PROVIDE PROTECTION
● ASSESS RISK OF FUNCTIONS	● EDUCATE STAFF ON RESPONSIBILITIES
● REPORT ON OR FOLLOW-UP ON RECOMMENDATIONS	● REPORT ON OR CORRECT PROTECTION DEFICIENCIES

the audit staff full authority and access to all of the organization's records, properties, and personnel relevant to the subject area under review. The audit staff usually also has the authority to review and appraise policies, procedures, and guidelines to ensure that they meet the requirements in the organization's policy statements. The head of the audit staff has full and independent access to all levels of management, as well as to the audit committee of the board of directors.

Since the audit staff has access to the various levels of management and to the audit committee, they can alert high-level management to security concerns. Additionally, they can validate your judgement and legitimize requests for staffing, equipment, and software. Audit findings can be used to highlight exposures and help set future priorities. Finally and most importantly, an extensive and complete audit can demonstrate the integrity of the security program.

BUILDING A SUPPORTIVE RELATIONSHIP

A security-audit committee is a good way to establish a dialogue between the security and audit staffs. The committee should meet at least semi-annually and should consist of the security coordinators (ISSO and security administrator) and the audit manager or managers. The goal of this committee would be to prioritize security activities and to negotiate those priorities. A meeting of this committee should precede the annual review of the security five-year plan.

After establishing a working relationship, this same committee could review and resolve specific department or unit exposures. The security and audit staff would be able to develop a consolidated business plan to address and resolve conscious breaches of organizational policy. As a result, the local units and departments will then be more willing to take advantage of the consulting role provided by these two staffs.

AUDIT PLAN

By establishing a working relationship with the audit staff, the security staff will have a say in the annual audit plan. The security staff will have prior knowledge of the directions, specific areas, and locations of the scheduled audits. By working together in the review process, the two staffs can ensure that the examination of the security process is complete, the findings are accurate, and that the unit responses are realistic and meet established organization requirements.

CONCLUSION

An organization's employees have always been the greatest threat to its efforts to maintain secure and confidential computer systems and information.

Numerous examples prove technology's inability to solve every security problem. An organization-wide program is needed to incorporate planning and security operations and should include the following:

1. A senior management totally committed to the goal of information security and willing to provide the leadership and act as an example to realize that goal.
2. The formulation of a set of simple and effective procedures and controls that are communicated regularly to every employee.
3. The systematic monitoring of the procedures by the organization's auditors and security officers to protect the organization's assets and quickly detect violations.
4. The willingness to investigate all violations and the resolution to identify training needs and to punish violators of the security program.

CHAPTER REVIEW

1. Compliance checking is the final element of a total security program.
2. All employees must have a thorough understanding of their responsibilities for compliance to all laws, regulations, policies, and procedures. Publication of an employee code of conduct may be an effective tool to create this awareness.
3. To assist in the compliance process, a self-administered internal audit program should be implemented.
4. Establish a positive, working relationship with your organization's audit staff.
5. The roles of the audit and security staffs are supportive; both are concerned with ensuring that the organization's assets are properly protected.

Exercise #1

Using the policy statement that you have created, develop an initial set of three to five questions that can be used as an internal audit questionnaire.

Exercise #2

Using the comparison of roles in Figure 20, identify other shared concerns that support the audit-security alliance.

CHAPTER 9

Computer and Information Security Laws

The proliferation of information, particularly in the computer industry, has made information more accessible and more vulnerable. At the same time, the general public has become more aware of computer control weaknesses as issues of personal privacy, computer fraud and legislation, and national security predominate in the news. The next section of this book will discuss the current laws affecting the security of computer systems and the information that is contained in them.

FEDERAL LAWS

Federal Antitrust Laws (Title 15 of the U.S. Code)

The exchange or sharing of competitively sensitive information with competitors about future product plans, marketing strategies, innovative manufacturing processes, and other conditions can lead to criminal (felony) and civil violations of the federal antitrust laws, resulting in fines up to one million dollars and treble civil damages. Individuals involved in such activities are subject to jail sentences of up to three years and fines up to one hundred thousand dollars.

Based on the language of the antitrust law, every organization is required to protect competitive information, keep processes secret from competitors, and resist efforts to obtain other organization's information.

Privacy Act of 1974

This act allows individual citizens to examine and make corrections to records maintained by the government.

Foreign Corrupt Practices Act (FCPA) of 1977 (Public Law 95-213) and 1988 Amendments (Public Law 100-418)

The FCPA requires companies to maintain books, records, and accounts that reflect, in reasonable detail, the disposition of the company's assets and to implement a system of internal accounting controls to safeguard the company's assets against unauthorized use or disposition.

The act applies to all corporations under the Securities and Exchange Commission (SEC) and is directed to senior managers and directors. The act established personal liability for noncompliance with fines up to ten thousand dollars and five years imprisonment. The act also fines corporations for non-compliance and requires that corporations annually certify to the SEC that they comply with the terms of the act. The act also permits civil suits from stockholders.

Federal Copyright Law (Title 17 of the U.S. Code)

The law makes it illegal to make or distribute copies of copyrighted material without authorization (section 106). One exception is that the user has the right to make a backup copy for archival purposes (section 117).

Unauthorized duplication of copyrighted material, including software, can lead to fines of up to one hundred thousand dollars and jail terms of up to five years per incident.

The doctrine of fair use is currently being tested in a U.S. District Court in New York. The complaint date was May 6, 1985, and the case is *American Geophysical Union et al* v. *Texaco Inc.* You should contact your legal staff to find the current status of this case and how it may affect your organization.

Computer Software Rental Amendments Act of 1990

This act was passed in the waning hours of the 101st Congress and states that the rental, lease, or lending of copyrighted software without the authorization of the copyright owner is expressly forbidden. This act does not apply to a nonprofit library or nonprofit educational institution.

The Counterfeit Access Device and Computer Fraud and Abuse Act of 1984 (Title 18 of the U.S. Code Chapter XXI)

This update of the USC makes it illegal to obtain unauthorized access to a government computer, or to access information held on behalf of the U.S. government.

Fines of as much as one hundred thousand dollars and twenty years in prison have been established.

Electronic Communications Privacy Act of 1986 (Public Law 99-508)

This law makes it illegal for persons to intercept telecommunications, such as electronic funds transfers and electronic mail, without prior authorization. However, in the case of *Dick Flanagan, et al.* v. *Epson America, Inc., et al.,* the court held that the entity providing the communication service is not liable for any offense regarding stored communications, including voice mail, e-mail, or other recorded communications.

Computer Fraud and Abuse Act of 1986 (Public Law 99-474)

The act establishes that anyone who knowingly accesses a computer without authorization or exceeds authorized access or causes losses totaling more than one thousand dollars or prevents authorized users from using the computer is considered to be in violation of U.S. federal law. Penalties include fines up to two hundred and fifty thousand dollars, five years in prison, and restitution of damage.

The act is restricted to "federal interest computers." These computers are not just those owned by the government or used by the government, but computers that access federal data, or computers that are located in two or more states. The definition would include a terminal in one state accessing a mainframe in another.

Computer Security Act of 1987 (Public Law 100-235)

This act identifies steps to be taken to improve the security and privacy of the information contained in federal computer systems. The act applies to all federal agencies, state agencies, and many government contractors. The act requires that:

1. A central authority be established to develop guidelines for protecting unclassified, but sensitive information stored in government computers.
2. Each agency formulate a computer security plan.
3. Each agency provide training for its employees on the threats and vulnerabilities of its computer systems.

It should be noted that in January, 1991, the National Institute of Standards and Technology and the National Security Agency set a deadline of 1992 for plans to jointly develop new computer security criteria for all agency procurements. The effort is intended to create a new federal information processing standard.

INTERNATIONAL SOFTWARE PROTECTION

Since 1989, the law firm of Fenwick & West has compiled a report called "International Legal Protection for Software" and made it available to the general public. The report is valuable to all multinational corporations and organizations attempting to understand the global implications of copyright protection for software products. Recent documents have discussed the European Economic Community Software Directive and the status of copyright protection in Eastern Europe. The report has also provided in-depth information on all aspects of copyright protection, various conventions, and a country-by-country breakdown of signatories.

To obtain a free copy of the current report, contact Fenwick & West at: 1920 N Street NW, #650, Washington, D.C. 20036; Telephone: (202) 463-6300; Fax: (202) 463-6520; Telex: 897136.

STATE LAWS

Forty-nine states have computer crime laws (as of April, 1991, only Vermont does not). Generally speaking, the individual acts make it illegal to attempt to gain unauthorized access to a computer system or to assist in unauthorized access to a computer system.

You should contact your legal staff for additional information or you can obtain the *Computer Crime Law Reporter* from the National Center for Computer Crime Data at (408) 475-4457.

INTERNATIONAL COMPUTER CRIME STATUTES

Australia

The Northern Territory Criminal Code Act (No. 47), 1983, section 276.
Crimes (Amendment) Ordinance (No. 4) No. 44, 1985, amending the New South Wales Crimes Act of 1900 in its application to the Australian Capital Territory, sections 93 and 115.

Canada

Canada Criminal Code, sections 342.1 and 430.

Federal Republic of Germany

Provisions of the penal code as amended by the Second Law for the Prevention of Economic Crimes of 1986, sections 202a, 263a, 269-274a, b, and c, and 348.

France

French Criminal Code, section 307-1-4.
Act No. 78-17 of January 6, 1978 on Data Processing, Data Files, and Individual Liberties, sections 41-44.

United Kingdom

The Forgery and Counterfeiting Act 1981.
Data Protection Act of 1984.
Computer Misuse Act of 1990.

Montreal Protocol

The Montreal Protocol was revised in June, 1990, phasing out chlorofluoro-carbons (CFC). In the computer security industry, this protocol directly affects the use of Halons 1211, 1301, and 2402. The following schedule identifies the phaseout agreement:

- January 1, 1992 — Freeze on new Halon installations.
- January 1, 1995 — 50 percent reduction.
- January 1, 2000 — 100 percent phaseout.

By January, 1993, any exceptions will be determined (such as submarine and airplane cockpits). All other Halon use must conform to the preceding phaseout schedule.

Due to the changes in computer and information laws, court cases, and government regulations, contact your legal staff for the current status of litigation in your location.

For researching international law, the books by James Arlin Cooper and Ulrich Sieber (listed in the Appendix) are recommended.

APPENDIX A: RESOURCES

ADDITIONAL READINGS

Ardis, Patrick M., and Comer, Michael J. *Risk Management: Computers, Fraud and Insurance.* London, 1987.

Bloombecker, Buck. *Spectacular Computer Crimes.* Homewood, IL, 1990.

Burger, Ralf. *Computer Viruses.* Grand Rapids, MI, 1988.

Cooper, James Arlin. *Computer & Communications Security: Strategies for the 1990s.* New York, 1989.

Gibson, Cyrus F., and Jackson, Barbara Band. *The Information Imperative.* Lexington, MA, 1987.

Grose, Vernon L. *Managing Risk.* Englewood Cliffs, NJ, 1987.

Mantz, Robert K., Mereten, Alan G., and Severance, Dennis G. "Senior Management Control of Computer Based Information Systems." Morristown, NJ, 1983.

Rochester, Jack B., and Gantz, William. *The Naked Computer.* New York, 1983.

Sieber, Ulrich. *The International Handbook on Computer Crime.* Chichester, England, 1986.

Stoll, Clifford. *The Cuckoo's Egg.* New York, 1989.

Vallabhaneni, S. Rao. *Auditing Computer Security: A Manual with Case Studies.* New York, 1989.

TRAINING VIDEOS

Commonwealth Films, Inc.
223 Commonwealth Ave.
Boston, MA 02116
(617) 262-5634
(617) 262-6948 (fax)

Data Security: Be Aware or Beware—This popular video uses a light, attention-getting style to dramatize computer security lapses in the office setting.

Locking the Door—Particularly effective for management and research personnel. Delivers the message that data, information, and hardware must be protected like any other asset. Useful in the PC and mainframe environments.

Invasion of the Data Snatchers—Designed for PC and mainframe users. Particularly effective for the clerical, operation, and administrative workforce, this award-winning video uses an entertaining, detective "comic strip" format to cover five hypothetical cases of computer security lapses.

Under Wraps: Information Security—Stresses the importance of preventing sensitive, proprietary, and confidential information of all kinds from leaking to people who have no need or right to know. Important do's and don'ts for safeguarding written, verbal, computer, and other information. Designed for all employees.

Mum's the Word: PC and LAN Security—This video focuses on data security for personal computers, laptops and Local Area Networks (LAN). Designed for all PC users—new hires, office and clerical personnel and management.

Maximum Security: Microcomputer Data & Hardware Security—Award-winning video for microcomputer and PC users. Situation comedy about Winslow Loman, a data security "repeat offender," delivers serious messages about microcomputer data and hardware protection and the consequences of breaches of security.

Lockout: Information Security—Fast-paced and as exciting as TV's "Jeopardy," the television game show video reviews and tests the viewer's knowledge of microcomputer protection practices for data and hardware, methods of classifying, labeling, and storing data and hardware.

Buried Alive: Document Retention—This video shows the legal, financial, and operational consequences for noncompliance in records management programs. Employees will identify with the scenario presented.

Write Now, Pay Later—Raises awareness of the legal risks of a casual attitude toward document writing. The video stresses the need for clear, conscious writing. An effective companion to *Buried Alive*.

PROFESSIONAL ORGANIZATIONS

ASIS—American Society for Industrial Security. 1655 North Port Meyer Drive, Suite 1200, Arlington, VA 22209. (703) 522-5800.

CSI—Computer Security Institute. 600 Harrison St., San Francisco, CA 94107. (415) 905-2267.

EDPAA—Electronic Data Processing Auditors Association. P.O. Box 88180, Carol Stream, IL 60188-0180. (312) 682-1200.

ISSA—Information Systems Security Association, P.O. Box 9457, Newport Beach, CA 92658. (714) 492-3462.

IIA—Institute of Internal Auditors. 249 Maitland Ave., Box 1119, Altamonte Springs, FL 32701. (305) 830-7600.

NCCCD—National Center for Computer Crime Data. 1222 17th Ave., Santa Cruz, CA 95062. (408) 475-4457.

APPENDIX B: DEFINITIONS

DATA CLASSIFICATIONS

Critical—Also used in the development of disaster recovery planning. Information or systems considered to be integral to the business and without which operations would be curtailed or otherwise severely impeded.

Essential—Information or systems less essential to the business than information or systems classified as Critical and without which operations would be difficult but not severely curtailed or impeded.

Internal-Use-Only—Information to be used by organization employees only, similar in scope and control to the Restricted classification.

Personnel/Private—Information regarding the employees. By law, physical, technical, and administrative safeguards must be implemented to ensure the security and confidentiality of the data. Normally this information is the responsibility of the personnel staff, but payroll and medical records are also included.

Privileged—Information concerning a lawyer-client relationship.

Public — Information that is not classified and is on public record, requiring only a normal business level of protection.

Restricted—Information that has not been classified as Secret or Confidential but has not yet been released to the public, or that is not intended for release outside the organization.

Sensitive—Information holding any classification level. Information classified as Secret, Confidential, Restricted, and Proprietary are all Sensitive classification levels.

Vital—Data elements in critical applications and systems that require appropriate backup and off-site storage.

KEY ROLES

Chief Executive Officer (CEO)—The CEO is responsible for adapting the use of computers to the business objectives of the company. To accomplish this, the CEO establishes policy, assigns responsibilities, allocates resources, and provides a system for measuring security.

Chief Financial Officer (CFO)—The CFO is responsible for ensuring the integrity of the company's accounting records and for the efficient use of all resources, including computers.

Director of Information Services (DIS)—The DIS is responsible for operation of the system, including supervision of system operators and any privileged users. The DIS is responsible for all the controls of and over the system.

Information Security Officer (ISO)—The ISO is responsible for developing the computer and information security policy to be adopted by senior management. The ISO is also responsible for advising on protective measures, including standards and procedures, measuring performance, and reporting to management. The ISO may supervise the system security administrator(s).

Senior Management—The senior manager of a business activity, such as the director of payroll, is responsible for the specification and operation of the controls for that activity. In addition, this individual is the owner of the company's data for that specific activity. The senior manager specifies the use of that data, including who may see and/or update it. Senior management may approve who, within the department, has access to the data and what transactions the employees may use. Because the senior management is responsible for the routine reconciliation of the department's activities, this position should not include the ability to originate data or transactions.

System Security Administrator (SSA)—The SSA is responsible for the creation and maintenance of the access control records. The SSA acts as surrogate for the system manager and the application and data owners. The SSA enrolls new users and grants access. The SSA works under the supervision of the director of information services or the ISO and is subject to review by the internal auditors. While the SSA may use electronic mail, the position should not include access to the system for any other purposes.

DATA PROCESSING

ACF2—Access Control Facility Version 2. Restricts data access for IBM mainframe computers.

CRT—Cathode ray tube. The screen found on terminals, consoles, and PCs. Also known as a video display terminal (VDT).

DES—Data Encryption Standard. An encryption algorithm published and supported by the National Institute of Standards and Technology (NIST).

Encryption—Transforming data to make it unreadable without the corresponding deciphering code.

Halon—A chorofluorocarbon (CFC) gas that has been installed in computer facilities to fight fires. Total phaseout scheduled for January, 2000.

LAN—Local area network. A system of interconnected devices, usually within a short distance of each unit.

NSA—National Security Agency.

Password—The authorization to allow access to a system or data by an individual.

PC—Personal computer designed to be used by a single user.

RACF—Resource Access Control Facility. IBM product that restricts access for IBM mainframe computers.

Security—Protection of organizational assets from harm. Security includes unauthorized access, modification, disclosure, or destruction—either accidental or intentional.

Software—All computer programs that control the operations of the hardware.

TSO—Time Sharing Option. Interactive mainframe computer processing.

Virus—An unauthorized program that copies itself into other programs whenever the trigger mechanism is executed.

INDEX